The Nothing Problem

Kris Michael McKenna

PP

Publishing Place

PP

Publishing Place

Published in the United States by Publishing Place LLC, Skowhegan, Maine, United States

www.udttheory.com

'The Nothing Problem' and the '*PP*' colophon are trademarks of Publishing Place LLC

ISBN: 978-1-969343-00-1 (*Paperback*)

ISBN: 978-1-969343-01-8 (*eBook*)

LCCN: 2025923287

Printed in the United States of America

First Edition 2025.

Publishing Place
55 Main Street
Skowhegan, ME 04976

www.publishingplace.net

CONTENTS

Introduction 1

An Ode to Gene 4

1. The Problem with Nothing 6
2. Why Historical Aether Failures Don't Apply 13
3. When Forces Become Flows 24
4. The Universal Equation 34
5. Gravity - The Gentle Giant 43
6. Electromagnetism - Nature's Whirlpools 53
7. The Nuclear Mysteries 65
8. The Vibrational Blueprint 76
9. Inertia - The Resistance Revealed 90
10. The Green Paint Problem 99
11. Faster Than Light (But Not Breaking Einstein?) 109
12. The Quantum Connection 118
13. Black Holes - The Ultimate Displacement 130
14. The Cosmic Crystal 144
15. The Eternal Cosmos 150
16. Engineering the Invisible 156
17. Rewriting the Textbooks 167

18. The Universe Revealed 179

Epilogue 188

The Nothing Problem - Comprehensive Glossary 189

The Nothing Problem - Conept Guide 197

Author's Note: Why I Wrote This 200

INTRODUCTION

This book is not a physics textbook. It contains no complex equations, no mathematical proofs, and no detailed experimental procedures. Instead, it's an invitation to think differently about the most fundamental questions in science: What is space? How do forces actually work? And why does modern physics, for all its mathematical success, leave so many of us feeling like we don't really understand how the universe operates?

For the past several years, I've been developing a theoretical framework called Unified Displacement Theory (UDT) — a mathematical model suggesting that all fundamental forces might emerge from mechanical processes in a physical medium that fills space. The technical papers, with their equations and statistical analyses, are available for researchers who want to examine the detailed mathematical foundations. But mathematics, however elegant, can only take us so far toward genuine understanding.

The ideas in this book grew from a simple observation that nagged at me throughout my limited physics education: we're told that space is "nothing," yet this nothing somehow has measurable properties, carries waves, and supports fields. We're asked to accept that forces act mysteriously across empty space, that particles can be in multiple places simultaneously, and that the deepest truths of physics are fundamentally incomprehensible to human intuition.

What if this incomprehensibility isn't a feature of nature, but a deterioration in our thinking?

What if physics became mysterious only when we stopped asking how things actually work and settled for mathematical descriptions of what we observe? What if the universe operates through mechanisms as understandable as any human engineering project, but we've trained ourselves not to look for them?

This book explores the possibility that space isn't empty, but filled with a subtle medium whose mechanical properties create everything we observe as fundamental physics. It's not anti-science — it's an attempt to restore the mechanical understanding that made science comprehensible in the first place. Every analogy, every visualization, every mechanical explanation in these pages is grounded in rigorous mathematical models, even though I've chosen to present the ideas through intuitive reasoning rather than formal equations.

My goal is to serve as a bridge: to help general readers explore a novel theory of how things work, while encouraging scientists to consider whether mechanical explanations might illuminate aspects of physics that mathematical abstractions alone cannot reach. If you're a researcher intrigued by these ideas, I invite you to examine the technical papers where these concepts are developed with full mathematical rigor. If you're simply someone who has always felt that physics should make intuitive sense, I invite you to explore what becomes possible when we stop accepting mystery as the price of knowledge.

We live in an age where the public has unprecedented faith in science, yet science itself has become increasingly incomprehensible to the very people who support and fund it. This creates a dangerous disconnect between scientific authority and public understanding. I believe we can do better. We can have physics that is both mathematically rigorous and intuitively comprehensible, both experimentally validated and mechanically satisfying.

The journey ahead takes us from a simple logical contradiction about the properties of "nothing" to a complete reimagining of how forces, matter, and even consciousness might work. Along the way, we'll discover that making physics comprehensible doesn't diminish its wonder — it reveals an elegance and beauty that pure mathematical abstraction can never quite capture.

The universe may indeed be comprehensible, as Einstein hoped. But comprehensibility requires more than mathematical formalism — it requires mechanisms we can visualize, processes we can understand, and explanations that satisfy our deepest need to know not just what happens, but how and why it happens.

The nothing problem is real, and it's time we solved it.

Kris Michael McKenna
2025

An Ode to Gene

A small tribute to Gene Rodenberry

Gene Roddenberry walked a path remarkably similar to my own - from law enforcement to visionary storytelling, from practical problem-solving to imagining how the universe might actually work. As a police officer turned writer and filmmaker, he understood that the most profound truths often emerge from the intersection of practical experience and boundless imagination.

What strikes me most about Roddenberry's vision is how it anticipated the very substrate mechanics that UDT now describes mathematically.

His vision of replicators weren't just convenient plot devices - they were explorations of matter reorganization through frequency patterns.

His transporters weren't simply faster travel - they were investigations into consciousness and pattern preservation.

His warp drive wasn't mere speed - it was a meditation on the nature of space itself.

Most remarkably, his character Q embodied what substrate mastery might actually look like: the ability to manipulate reality through direct interface with the organizing principles of the universe. Q's seemingly omnipotent abilities make perfect sense as advanced substrate manipulation - reorganizing matter through thought, traveling instantaneously through substrate space, and even manipulating time through substrate flow modifications.

Roddenberry intuited that technology and consciousness would eventually merge, that matter and energy and information were aspects of the same underlying reality, and that sufficiently advanced civilizations would transcend the limitations that seem absolute to us. These may have been fantasies - but testaments of extrapolations of substrate physics principles we're only now beginning to understand mathematically.

Like Roddenberry, I've found that real-world experience in investigation and analysis provides crucial grounding for theoretical exploration. The same skills that solve cases can illuminate cosmic mysteries. The same curiosity that asks "how did this crime unfold?" can ask "how does reality actually work?"

Roddenberry showed us a future where humanity had learned to work with the fundamental organizing principles of reality rather than against them. Where technology served consciousness rather than dominating it. Where understanding the deeper mechanics of existence led not to cold materialism, but to greater wisdom, compassion, and wonder.

If UDT proves correct, then Star Trek wasn't science fiction - it was substrate physics intuition. Gene Roddenberry may have been humanity's first substrate theorist, exploring through fiction what mathematics is only now catching up to describe and has in the past.

The universe he imagined - where matter reorganizes through frequency, where consciousness networks across cosmic distances, where reality responds to understanding rather than force - may be exactly the substrate universe we actually inhabit.

Thank you, Gene, for showing us not just what technology might become, but what we might become when we finally understand that we're swimming in an ocean of infinite possibility.

CHAPTER 1
THE PROBLEM WITH NOTHING
WHY PHYSICS GOT LOST IN ITS OWN EQUATIONS

Picture yourself sitting in your favorite chair, reading this book. You feel solid, grounded, completely at rest. But according to modern physics, you're actually hurtling through absolute nothingness at tremendous speed - the Earth spinning at 1,000 miles per hour, orbiting the Sun at 67,000 miles per hour, the entire solar system racing around the galaxy at 514,000 miles per hour.

And through all this cosmic motion, you remain perfectly comfortable in your chair because... well, that's where things get interesting. Or rather, where they seem impossible.

The Invisible Contradiction

Modern physics has a secret embarrassment, a logical contradiction so fundamental that physicists have simply learned to look the other way. It's called the Nothing Problem, and once you see it, you can't unsee it.

The problem is this: Physics tells us that space is nothing - literally empty vacuum, devoid of any material content. Yet this same "nothing" somehow manages to:

Carry waves at exactly 186,282 miles per second.

Store vast amounts of energy in its vacuum fluctuations.

Bend, stretch, and curve in response to matter.

Support magnetic fields that can lift trains.

Transmit gravitational effects across billions of miles.

Stop and think about this for a moment. How can "nothing" do anything at all?

It's like saying an empty swimming pool can carry waves, or that a non-existent rope can pull objects together. The contradiction is so obvious that it seems almost childish to point out, yet this is precisely how our most advanced theories describe reality.

When Nothing Has Properties

Imagine trying to explain to a curious child why their toy magnet works. You might say something like: "The magnet creates an invisible field that reaches out through empty space and pulls on the metal."

But the child, with the clarity that adults have trained themselves to ignore, asks the obvious question: "If space is empty, how does the invisible field get from the magnet to the metal?"

You might answer: "Well, the field just... travels through space."

"But you said space was empty. How can something travel through nothing?"

"The field doesn't need anything to travel through. It just... propagates."

"What does 'propagates' mean?"

And there you are, trying to explain how something moves through nothing, supported by nothing, to affect something else. The child's confusion is perfectly justified - you're describing something impossible.

The Properties of "Nothing"

Let's catalog what this "nothing" supposedly does:

It has a speed limit. Light travels through vacuum at exactly 983,571,056 feet per second - not faster, not slower. This suggests vacuum has a specific property, like the speed of sound through air or water. But air and water are something. Vacuum is supposed to be nothing.

It has electrical properties. Space has a measured "permittivity" - essentially its ability to support electric fields. It also has "permeability" - its response to magnetic fields. These aren't just abstract mathematical concepts; they're measurable properties with specific values. Properties belong to things, not to nothing.

It has energy. Quantum mechanics tells us that empty space seethes with energy fluctuations. The Casimir effect demonstrates this experimentally - two metal plates placed close together in vacuum experience a measurable force because the space between them contains less energy than the space outside. But how can nothing contain energy?

It can curve. Einstein showed us that massive objects curve spacetime, and this curvature is what we experience as gravity. But curvature is a geometric property. Geometry requires something to be geometric about. You can curve a rubber sheet or spacetime fabric, but you cannot curve pure nothingness.

The Historical Mistake

This contradiction didn't arise because physicists are illogical. It emerged from a historical misunderstanding that became so deeply embedded in physics that questioning it feels almost heretical.

A century ago, physicists did believe space was filled with something - the "aether," an invisible medium that carried light waves just as air carries sound waves. This made perfect sense mechanically. Waves need something to wave in.

But then came the famous Michelson-Morley experiment, which failed to detect Earth's motion through this supposed aether. The conclusion seemed clear: there is no aether. Space is empty.

However, what the experiment actually disproved was a *static* aether - a fixed, immovable reference frame. The possibility of a dynamic, flowing medium that moves with matter was never ruled out. But by then, the mathematical formalism of field theory had become so successful that physicists stopped asking what fields actually *are*. They became content with mathematical descriptions that worked, even if they made no physical sense.

Einstein himself later regretted this turn. In a 1920 speech at the University of Leiden, he said: "According to the general theory of relativity, space without aether is unthinkable; for in such space there would be no propagation of light, but also no possibility of existence of measuring rods and clocks."

But by then, it was too late. Physics had committed to the impossible: properties without substance, effects without medium, action at a distance through absolute nothingness.

The Intuitive Alternative

What if space isn't empty? What if it's filled with something so subtle, so perfectly transparent to matter, that we've simply never noticed it directly?

Think about a fish trying to understand water. A fish has never experienced "not-water," so water might seem like nothing to it. Only by studying how things move through water - how pressure changes, how currents flow, how waves propagate - might the fish deduce that it lives in a medium with specific properties.

We might be in a similar situation. What we call empty space might actually be filled with an incredibly subtle medium - call it the substrate - that interacts with matter in ways we're only beginning to understand.

If this substrate exists, then suddenly physics becomes intuitive again:

Gravity - isn't a mysterious force acting across empty space - it's pressure differences in the substrate, like how low pressure systems in the atmosphere create wind. How pressure in the ocean creates acceleration through restorative force.

Electromagnetic fields - are not mystical entities - they're flow patterns in the substrate, like whirlpools in water.

Light waves - are not self-propagating disturbances in nothing - they're actual waves in an actual medium, just like sound waves in air.

The Revolutionary Possibility

This book explores a theoretical framework called Unified Displacement Theory (UDT), which takes seriously the possibility that space is not empty but filled with a dynamic substrate that makes physics work. This isn't just philosophical speculation - it's a mathematical framework that may explain how all the fundamental forces actually operate through a single, comprehensible mechanism.

The implications, if correct, would be revolutionary:

Physics becomes mechanistic and intuitive rather than mysterious and mathematical.

All fundamental forces unify under a single principle.

New technological possibilities emerge, from gravity control to faster-than-light communication.

The universe transforms from a collection of bizarre quantum mysteries into an elegant, understandable machine.

But we're getting ahead of ourselves. First, let's explore what happens when we stop pretending that nothing can have properties, and start asking what that something might actually be.

The journey begins with a simple recognition: If it acts like something, stores like something, and affects like something, then maybe - just maybe - it *is* something.

Welcome to the substrate universe. Physics is about to become intuitive again.

"According to the general theory of relativity, space without aether is unthinkable; for in such space there would be no propagation of light, but also no possibility of existence of measuring rods and clocks."

Einstein (1920)

CHAPTER 2

WHY HISTORICAL AETHER FAILURES DON'T APPLY

LEARNING FROM PAST MISTAKES WITHOUT REPEATING THEM

If you've had any physics education, alarm bells are probably ringing right now. "Wait a minute," you might be thinking, "this substrate idea sounds suspiciously like the old aether theory that was thoroughly debunked over a century ago. Didn't the Michelson-Morley experiment prove there's no medium filling space?"

This is a perfectly reasonable concern, and it deserves a careful answer. The short version is this: yes, 19th-century aether theories failed spectacularly, but they failed for very specific reasons that don't apply to modern substrate theories. Understanding why those theories failed - and how substrate theory avoids their fatal flaws - is crucial for evaluating whether the substrate model deserves serious consideration.

The relationship between historical aether theories and modern substrate theory is like the relationship between bloodletting and modern medicine. Both involve the human body, but the mechanisms, understanding, and methods are so fundamentally different that the failure of one tells us nothing about the validity of the other.

What the 19th Century Actually Tested

The aether theories that dominated 19th-century physics made very specific claims about how a light-carrying medium should behave. Understanding exactly what those theories predicted - and what experiments were designed to test - is key to understanding why their failure doesn't invalidate substrate theory.

The 19th-Century Aether Model:

Static medium: The aether was assumed to be stationary in space, like a cosmic ocean that Earth moves through.

Detectable relative motion: Earth's movement through this stationary aether should create measurable effects, like feeling wind when you stick your hand out of a moving car.

Partial dragging: Some theories proposed that matter partially drags the aether along, but not completely.

Velocity-based effects: All predicted phenomena involved relative velocities between matter and the aether medium.

What Experiments Were Designed to Detect:

Aether wind: Measurable differences in light speed depending on Earth's direction of motion through the supposed stationary aether.

Seasonal variations: Changes in these effects as Earth orbits the Sun at different velocities.

Drag coefficients: Specific ratios of how much matter drags the aether along with it.

The crucial point is that every 19th-century aether experiment was looking for velocity differences between Earth and a medium. When they found no such differences, they concluded no medium exists.

The Michelson-Morley Experiment: What It Actually Showed

The 1887 Michelson-Morley experiment is often cited as the definitive refutation of medium-based theories, but this interpretation rests on a crucial misunderstanding of what the experiment actually demonstrated.

What the Experiment Tested: Albert Michelson and Edward Morley built an incredibly sensitive interferometer designed to detect Earth's motion through a stationary aether. They reasoned that if Earth is moving through a stationary medium at 30 kilometers per second (our orbital speed), light traveling in different directions should show measurable speed differences, just like sound waves in air show different speeds depending on wind direction.

What They Found: No detectable difference in light speed in any direction. The "aether wind" that should have been obvious if we were moving through a stationary medium was completely absent.

The Standard Interpretation: "Therefore, there is no aether. Space is empty."

The Alternative Interpretation: "Therefore, there is no relative motion between Earth and the local medium."

This second interpretation is crucial. The experiment showed no relative motion between Earth and whatever medium might exist locally. This is perfectly consistent with a medium that moves completely with matter - exactly what substrate theory proposes.

The Fatal Flaw in Historical Thinking

Imagine two passengers sitting in cars traveling side by side at exactly the same speed. If one passenger looks at the other and sees no relative motion, would it be reasonable to conclude that neither car is moving?

Of course not. The absence of relative motion between the cars tells us nothing about whether they're both moving relative to the road. Yet this is essentially the logical error that led to the dismissal of medium-based theories.

The Michelson-Morley experiment showed no relative motion between Earth and the local medium. The standard interpretation was: "No relative motion means no medium." But an equally valid interpretation is: "No relative motion means the medium moves completely with Earth."

The Historical Mistake: 19th-century physicists were so focused on finding relative motion between matter and medium that when they found none, they concluded the medium didn't exist. But complete co-movement between matter and medium was never properly considered as an alternative explanation.

Why Substrate Theory Is Different

Modern substrate theory makes fundamentally different claims that sidestep the specific problems that killed historical aether models:

Complete Local Co-Movement: Instead of assuming a static aether or partial dragging, substrate theory proposes that the substrate moves completely with matter at local scales. Earth and its immediate environment are carried along by substrate flow, just like passengers in a smoothly moving train feel no motion relative to the train itself.

Pressure-Based Effects: Instead of looking for velocity differences, substrate theory predicts pressure-gradient effects. The substrate's mechanical properties create forces through density variations and flow patterns, not through relative motion between matter and medium.

Scale-Dependent Coupling: Different substrate responses occur at planetary scales versus laboratory scales. Effects that are undetectable in small laboratory experiments might be significant across cosmic distances.

Different Experimental Signatures: Substrate theory predicts effects in completely different physical domains than historical aether theories:

GPS timing variations from altitude-dependent substrate density.

Substrate pressure gradient effects on particle interactions.

Electromagnetic coupling variations with local substrate properties.

Long-range correlations maintained by substrate wave propagation.

These are not the velocity-based effects that 19th-century experiments were designed to detect.

The Technology Gap

An often-overlooked factor in dismissing medium-based theories is the enormous technological advancement since 1887. The sensitivity improvements in measurement technology over the past 140 years are almost incomprehensible.

1887 Michelson-Morley Technology:

Human eye observations through telescope eyepieces.

Sodium flame lamps (unstable, incoherent light).

11-meter path lengths with single-pass light.

Sensitivity: about 0.01 fringe shifts.

Manual measurements recorded by hand.

Modern LIGO-Class Technology:

Quantum-limited photodetectors with digital processing.

Ultra-stable laser systems (coherent, single-frequency).

4-kilometer arms with 300-bounce light amplification.

Sensitivity: 10^{-21} strain detection (1/1000th the width of a proton).

Automated data acquisition and computer analysis.

The Staggering Improvement: Modern interferometry is approximately **one billion times** more sensitive than the 1887 experiment. Effects that were completely undetectable in Michelson and Morley's time are now easily measurable.

Even more intriguing, Michelson and Morley did observe a small "residual" effect of about 0.02 fringes that they dismissed as experimental error. With modern precision, what appeared as noise in 1887 might actually contain genuine substrate signatures that can now be isolated and analyzed.

The Molecular Parallel

The dismissal of substrate theories based on 19th-century experiments is remarkably similar to what could have happened with molecular theory. In the 1800s, scientists had perfectly working thermodynamic laws but no direct evidence for molecules. If someone had concluded "since we can't detect individual molecules, they don't exist, and matter must be continuous," they would have been making the same logical error.

Just as molecules existed below the detection threshold of 19th-century technology, substrate effects might exist below the threshold of 19th-century aether experiments while being perfectly detectable with modern precision.

Historical Precedent: This pattern repeats throughout science.

Radio waves existed before radio receivers could detect them.

Bacteria existed before microscopes could reveal them.

Gravitational waves existed for a century before LIGO could measure them.

The absence of detection doesn't prove absence of phenomenon - it often just means the technology isn't ready yet.

Different Physics, Different Tests

The most important point is that substrate theory and historical aether theory propose fundamentally different physical mechanisms. Rejecting substrate mechanics because aether dragging failed is like rejecting molecular kinetic theory because caloric fluid theory failed - they involve different underlying physics despite both dealing with similar domains.

Historical Aether Predictions:

Measurable relative velocities between Earth and medium.

Light speed variations depend on Earth's orbital motion.

Specific drag coefficients for different materials.

Seasonal variations in optical experiments.

Modern Substrate Predictions:

GPS timing variations from gravitational substrate density changes.

Particle accelerator signatures at specific energy levels.

Electromagnetic coupling variations with substrate properties.

Faster-than-light information transfer through substrate waves.

These predict effects in completely different experimental domains. The 19th-century experiments were testing for the wrong physical signatures to evaluate substrate theory.

The Scientific Method Applied Correctly

The scientific method requires testing the actual claims of a theory, not refuting different claims made by superficially similar theories. Historical aether experiments tested specific velocity-based mechanisms. Modern substrate theory proposes pressure-based mechanisms.

The Analogy: If you wanted to test whether a car engine works by examining the horse that used to pull the carriage, your failure to find a horse wouldn't tell you anything about the engine. The car and the horse-drawn carriage both provide transportation, but the mechanisms are so different that tests designed for one tell you nothing about the other.

Similarly, substrate theory and historical aether theory both propose that space contains a medium, but the proposed mechanisms are so different that tests designed for historical aether tell us nothing about modern substrate theory.

What This Means for Substrate Theory

The failure of 19th-century aether theories doesn't invalidate substrate theory any more than the failure of geocentric astronomy invalidated the study of planetary motion. Both involved careful observation and mathematical analysis of celestial phenomena, but the underlying models were fundamentally different.

Substrate theory deserves evaluation based on:

Its own specific predictions.

Experiments designed to test its actual claims.

Modern technology capable of detecting subtle effects.

Statistical analysis of contemporary high-precision data.

Not based on:

The failure of theories with different mechanisms.

19th-century technology limitations.

Guilt by superficial association with discredited models.

The Path Forward

Understanding the distinction is crucial for fair scientific evaluation. Just as thermodynamics worked perfectly for 200 years before anyone understood the molecular mechanisms underneath, Einstein's mathematical formalism might describe emergent effects of substrate mechanics that 19th-century physicists couldn't detect because they were looking for the wrong signatures.

The substrate model makes specific, testable predictions that can be evaluated with modern experimental techniques. These predictions should be tested rigorously and objectively, without prejudice based on the failure of fundamentally different theories from over a century ago.

Science advances by building on past knowledge while avoiding past mistakes. The mistake of 19th-century aether theory was assuming a static medium and looking for velocity effects. The wisdom we can extract from that failure is to consider dynamic media and look for pressure effects.

That's exactly what substrate theory does.

A Case Reopened

Historical aether theories failed because they made wrong assumptions about how a space-filling medium should behave.

They assumed:

Static or partially dragged medium.

Detectable relative motion with matter.

Velocity-based observable effects.

Modern substrate theory assumes:

Completely co-moving dynamic medium.

No relative motion with local matter.

Pressure-gradient observable effects.

These are fundamentally different physical models that should be evaluated independently. The failure of the first tells us nothing about the validity of the second, just as the failure of steam-powered cars tells us nothing about the viability of electric vehicles.

Both involve mechanized transportation, but the mechanisms are so different that the success or failure of one is irrelevant to the other. The substrate model deserves evaluation based on its own merits and predictions, not dismissal based on guilt by superficial association with discredited theories from a different era of physics.

"Any particle, ever isolated, has to be imagined as in continuous 'energetic contact' with a hidden medium."

Louis de Broglie

Chapter 3

When Forces Become Flows

Discovering the Hidden Machinery of Reality

Picture a leaf floating on the surface of a gently flowing stream. The leaf doesn't 'know' about the water beneath it - from the leaf's perspective, it's simply moving through space. But we can see the real story: the leaf moves because water is flowing, creating pressure differences that carry it along.

Now imagine that you are the leaf, and gravity is the stream. You feel yourself falling, but you can't see what's actually moving you...

What just happened? Ask most people, and they'll say "gravity pulled you down." Ask a physicist, and they might explain that you followed the curved geometry of spacetime, or that you moved along the path of least action through a gravitational field.

The Scale That Changes Everything

The Extraordinary Number: 10^{60} substrate granules fit inside a single atomic nucleus.

This isn't just big—it's incomprehensibly vast. If each granule were a grain of sand, you'd need more sand than exists on every beach, in every desert, on every planet in a billion galaxies to fill just one nucleus.

But this almost infinite granularity isn't just a curiosity. It's the key to understanding why one substrate medium can create all fundamental forces.

Why This Scale Works

Nuclear diameter: ~10^{-15} meters

Granule size: ~10^{-35} meters (Planck scale)

Granules per nucleus: ~10^{60}

The genius of 10^{60} granules per nucleus:

Fine enough to interact directly with quarks and electrons.

Numerous enough to behave as a smooth, continuous fluid.

Universal enough to fill all space uniformly.

Responsive enough to create different effects at different scales.

The Scale Hierarchy

Nuclear forces: Complete granule displacement creates extreme pressure gradients.

Electromagnetic forces: Granule stirring creates circulation patterns.

Gravitational forces: Gentle pressure waves through the granule sea.

Inertial forces: Granule compression resistance during acceleration.

Why 10^{60} Is Perfect

Too few granules (say, 10^{6}):

Substrate would be lumpy and irregular.

Forces would be choppy at different scales.

No smooth transitions between regimes.

Just right (10^{60} granules):

Perfect fluid behavior at all observable scales.

Statistical smoothness - even dramatic changes appear continuous.

Seamless scaling from nuclear to galactic.

Universal medium that enables all force types.

The Invisibility Factor

With 10^{60} granules per nucleus, the substrate is **too fine to detect directly**. It's like trying to see individual water molecules while swimming in the ocean. Traditional physics missed this because "empty" space appears smooth—but it's actually an incomprehensibly dense granule sea.

Every cubic centimeter of space contains 10^{90} substrate granules - more components than atoms in billions of galaxies, all working together to create the forces that hold reality together.

10^{60} substrate granules per nucleus makes the impossible possible: one medium, all forces, unified physics.

Energy Bubbles: Tessellated Substrate Granules

What do those 10^{60} granules actually look like? The simplest and most powerful picture is this: they are **energy bubbles**, spherical packets of structured energy that fill space and lock together when pushed.

Alone in free space each granule prefers a spherical shape (the minimum-energy form). Push many of them together and they naturally tessellate into polyhedral cells — a kind of energy foam. This is not a metaphor. The same geometric inevitability we see in soap bubbles and dense packings applies: spheres under compression transform into tightly-fitting polyhedra, producing a mechanically stable lattice that behaves like a continuous medium at larger scales.

Key properties of these energy bubbles:

No rest mass, only energy density. Each granule contains a fixed packet of energy; it has inertia through its interaction with neighbors, not because it carries mass itself.

Surface cohesion. Boundaries between bubbles represent steep energy-density gradients; deforming a boundary costs energy, producing an analogue of surface tension.

Quantized content and uniform size. Quantum rules lock each bubble's energy content, which stabilizes the tessellation and prevents runaway coalescence.

Fluid, yet structural. The foam flows and transmits pressure waves like a fluid, but the tessellation locks local geometry so it behaves mechanically like an elastic medium when stressed.

Why this picture matters: it gives us a concrete mechanism for every force. Gravity becomes long-wavelength compression in the foam; electromagnetism becomes circulating shear and vortex patterns in the cells; nuclear binding is what happens when many bubbles are forcibly compressed into a tiny region, the foam pushes back with enormous short-range pressure. Inertia is simply the foam's resistance to changing its local packing when matter accelerates.

This tessellated view explains several otherwise puzzling features at once:

Density without mass: space possesses effective density from energy per volume.

Stability at extreme scales: tessellation plus quantization prevents the substrate from smearing out, even down to nuclear scales.

Preferred geometries and resonances: the polyhedral packing imposes geometric constraints that can produce discrete, testable resonances and directional effects.

Surface phenomena: interactions that depend on boundary area (rather than only bulk mass) naturally follow from energy-boundary physics.

A few practical consequences to keep in mind: the bubbles' characteristic size sets the bridge between the microscopic quantum world and the smooth macroscopic world we observe. Pin down that scale — whether rooted in the Planck length, a fundamental energy quantum, or a cosmological density constant and you anchor the whole model. Likewise, matter can be modeled as topological defects, cavities, or tightly bound knot-like configurations within the foam; that geometric coupling is what yields discrete particles with charge, spin, and mass-energy.

In short: rather than empty space or a purely abstract field, imagine an invisible, tessellated energy foam. Forces are just different modes of straining, flowing, and compressing that foam. Once you accept that picture, the behavior of gravity, electromagnetism, and the strong force becomes a single set of fluid-mechanical phenomena expressed through geometry and quantized energy packets.

The Pressure Revolution

What we call "gravity" might actually be pressure differences in this substrate, pushing objects together rather than mysteriously pulling them. What we call "electromagnetic fields" might be circulation patterns in the substrate, like invisible whirlpools that guide charged particles. What we call "nuclear forces" might be extreme compression zones where the substrate behaves like a fluid under tremendous pressure.

If this sounds far-fetched, consider this: every force we've ever discovered operates exactly like we'd expect forces to work in a flowing medium.

Gravity as Flow

Let's start with gravity, the most familiar force. Isaac Newton described it as objects mysteriously attracting each other across empty space - action at a distance. Einstein improved on this by describing it as curved spacetime, but that still leaves us with the puzzle of how matter "knows" to follow these curves. But what if gravity works more like a fluid?

Imagine the substrate as an invisible fluid filling all of space. When matter exists within this fluid, it creates cavitation zones - areas of reduced substrate density within and throughout the matter itself. The substrate pressure at the boundary of this matter creates a restorative force, pushing objects with it to get back to a normal state. This boundary pressure differential is what we experience as gravitational acceleration.

Picture a bowling ball placed on a stretched rubber sheet. The ball creates a depression, and if you roll marbles nearby, they curve toward the ball. But here's the key insight: the marbles aren't being "attracted" to the ball - they're being pushed by the slope of the sheet itself.

Now imagine this happening in three dimensions, with the substrate playing the role of the rubber sheet. Massive objects like planets and stars create three-dimensional cavitation zones in the substrate - internal regions where substrate density is lower, while maintaining higher pressure at the boundaries. Objects are squeezed toward these displacement zones by the pressure differential. Other objects are naturally squeezed toward these displacement zones by the higher substrate pressure at their boundaries. The substrate pressure differential pushes objects from high pressure regions toward the internal cavitation zones where matter has displaced the substrate.

This isn't just a pretty analogy - it's a precise mathematical description. The acceleration of any object in this substrate field follows a simple equation:

[acceleration = -(1/substrate_density) × pressure_gradient]

This single equation might explain everything from why your coffee cup stays on the table to why galaxies spiral in space.

Electromagnetic Whirlpools

Electromagnetic forces present an even more striking example of substrate mechanics in action. Have you ever watched iron filings arrange themselves around a magnet? They form beautiful curved patterns that seem to reveal invisible "field lines." But what if those patterns aren't just revealing mathematical abstractions - what if they're showing us the actual flow structure of the substrate itself?

Picture the substrate as being capable of circulation, like water forming whirlpools. When electric charge moves through the substrate, it creates persistent circulation patterns - invisible vortices that extend outward through space. These circulation patterns are what we call magnetic fields.

Other charges moving through these circulation zones experience forces - not because of mysterious "field interactions," but because they're literally swimming through the moving substrate. It's like trying to swim across a river with a strong current; you get pushed sideways by the flow.

This explains why moving electric charges create magnetic fields (they stir up circulation in the substrate) and why changing magnetic fields create electric forces (changing circulation patterns create pressure waves that push on charges). It's all fluid mechanics, playing out in a medium we've never directly perceived.

The Nuclear Pressure Cooker

At the smallest scales, where protons and neutrons bind together in atomic nuclei, the forces become so strong that they've mystified physicists for decades. How can particles stick together with forces thousands of times stronger than electricity, but only across incredibly tiny distances?

The substrate model suggests a startling answer: nuclear forces might arise when multiple particles create such intense displacement in the substrate that it behaves like a fluid under extreme compression.

Imagine several marbles pressed tightly together in a container of thick honey. The honey around them gets squeezed into a smaller space, creating enormous pressure. This compressed honey would push back on the marbles with tremendous force, holding them tightly together. But this force would only operate across very short distances - just within the compressed zone.

This is essentially what might happen in atomic nuclei. Multiple particles clustered together create zones where the substrate becomes so compressed that it generates enormous inward pressure, binding the nucleus together. The "strong nuclear force" isn't a separate force at all - it's just the substrate's response to extreme compression.

The Universal Pattern

Notice the pattern emerging here. Every fundamental force seems to operate through the same basic mechanism:

Gravity: Gentle pressure gradients in the substrate.

Electromagnetism: Circulation patterns in the substrate.

Nuclear forces: Extreme compression zones in the substrate.

Instead of four separate, mysterious forces, we might have one simple, understandable mechanism operating at different intensities. The substrate responds to matter in predictable ways, and these responses are what we experience as the fundamental forces of nature.

Seeing the Invisible Ocean

This perspective transforms how we see the universe. Instead of objects floating in empty space, mysteriously affecting each other through invisible fields, we have objects swimming in a dynamic medium, pushing and flowing and circulating in ways that create all the phenomena we observe.

You're not sitting in empty space reading this book. You're sitting at the bottom of an invisible ocean - an ocean of substrate that extends from your chair to the farthest galaxies. Every movement you make sends pressure waves through this ocean. Every heartbeat creates tiny circulation patterns. Every thought (which involves electrical activity in your brain) stirs the substrate in infinitesimal but real ways.

The Mechanical Universe Returns

This isn't just about replacing one mathematical description with another. It's about returning to a mechanical understanding of the universe - one where effects have clear, visualizable causes, where forces operate through contact rather than action at a distance, where the universe becomes comprehensible rather than mysterious.

In this view, physics isn't about memorizing abstract field equations or accepting quantum weirdness as fundamental. It's about understanding how an incredibly subtle fluid medium responds to the presence and motion of matter. It's about recognizing that we live not in empty space, but in the most elegant fluid dynamic system ever discovered.

The implications stretch far beyond philosophical satisfaction. If forces are really flowing in a substrate medium, then we might learn to control those flows. Just as we learned to harness electromagnetic forces once we understood them, we might learn to manipulate substrate flows to achieve what seems impossible today: gravity control, faster-than-light communication, and energy systems that work with the substrate rather than against it. But first, we need to understand exactly how this substrate might work, and why a single equation might explain everything. That story begins with the most successful equation in the history of physics - one that might describe the hidden mechanics of reality itself.

"That gravity should be innate, inherent, and essential to matter, so that one body may act upon another at a distance through a vacuum, without the mediation of anything else... is to me so great an absurdity that I believe no man who has in philosophical matters a competent faculty of thinking can ever fall into it."

Newton

CHAPTER 4

THE UNIVERSAL EQUATION

ONE FORMULA TO RULE THEM ALL

Every great breakthrough in physics can be traced back to a single moment when someone saw a pattern that had been hiding in plain sight. Newton watching an apple fall and realizing it followed the same rules as planetary motion. Einstein recognized that space and time were aspects of the same thing. Maxwell discovered that electricity and magnetism were two faces of a single phenomenon.

What if the next great breakthrough is equally simple? What if every force in the universe - from the gentlest gravitational tug to the most violent nuclear explosion - follows the same basic rule?

The Search for Unity

For over a century, physicists have dreamed of a "theory of everything" - a single equation that would explain all the fundamental forces. They've constructed elaborate mathematical frameworks with names like supersymmetry, string theory, and loop quantum gravity. These theories require eleven dimensions, particles that exist only in mathematical abstractions, and energy scales that dwarf anything we could ever hope to observe.

But what if they've been looking in the wrong direction? What if the unifying equation isn't incredibly complex, but elegantly simple?

The Revelation

That one equation we referenced earlier that might change everything looks almost disappointingly ordinary:

[acceleration = -(1/substrate_density) × pressure_gradient]

Or in the mathematical shorthand that physicists prefer:

$$[a = -(1/\rho_{QS}) \times \nabla P_{QS}]$$

That's it. One equation. No exotic dimensions, no imaginary particles, no energies beyond comprehension. Just a statement about how things accelerate in response to pressure differences in a medium.

But this simple equation might explain:

Why your coffee cup doesn't float away (gravity).

Why magnets work (electromagnetism).

Why atomic nuclei hold together (strong nuclear force).

Why some particles decay (weak nuclear force).

Why you feel pushed back in an accelerating car (inertia).

All from pressure gradients in an invisible substrate that unify all known forces.

Gravity: The Gentle Pressure

When you drop a coin, Earth's mass has displaced substrate throughout its volume, creating zones of reduced substrate density. The higher substrate pressure at Earth's boundary creates a pressure differential that squeezes objects downward toward the displacement zone. This boundary pressure effect weakens with altitude, which is why gravity decreases as you move away from Earth's surface.

The acceleration you observe (about 32.2 feet per second squared at Earth's surface) comes directly from dividing the pressure gradient by the substrate density. The remarkable thing is that this calculation works not just for Earth, but for every massive object in the universe.

Consider our solar system. The Sun, with its enormous mass, displaces substrate throughout its volume, creating internal zones of reduced substrate density. The higher substrate pressure at the boundaries continuously squeezes planets toward the Sun's effective displacement zone. This boundary pressure effect maintains orbital mechanics - planets aren't falling 'downhill' toward the Sun, they're being squeezed inward by the substrate pressure differential while their orbital velocity keeps them circling around the pressure gradient.

The equation explains why Mercury orbits faster than Earth (it's deeper in the Sun's pressure gradient), why Pluto takes so long to complete an orbit (it's barely feeling the pressure difference), and why comets follow such elliptical paths (they're riding the pressure contours like a roller coaster).

Electromagnetism: Pressure in Motion

The same equation explains electromagnetic forces, but now we're dealing with moving pressure gradients - circulation patterns in the substrate that we call magnetic fields, and pressure waves that we call electric fields.

When you rub a balloon on your hair and it sticks to the wall, the equation describes how charge differences create substrate circulation patterns that generate pressure differentials, with higher pressure regions squeezing the balloon toward areas where the substrate circulation creates lower pressure zones near the wall. When you move a magnet near a piece of iron, the equation explains how the circulation patterns in the substrate push on the iron's internal charge structures.

The beauty is that the electromagnetic calculations use exactly the same equation as gravity, just with different values for the pressure gradients. The substrate responds to charge movement the same way it responds to mass - by creating pressure differences that push objects around.

Nuclear Forces: Extreme Pressure

At the smallest scales, where protons and neutrons cluster in atomic nuclei, the equation takes on dramatic proportions. Here, the substrate pressure gradients become so intense that they dwarf anything we experience in everyday life.

Multiple particles packed into incredibly tiny volumes create extreme substrate displacement - imagine trying to squeeze several bowling balls into a space the size of a marble. The substrate responds by generating enormous pressure gradients that hold the nucleus together with forces thousands of times stronger than electromagnetic repulsion.

The same equation that explains why planets orbit also explains why atomic nuclei don't fly apart, why nuclear fusion releases so much energy, and why nuclear fission can power entire cities. It's all pressure gradients in the substrate, just at different scales and intensities.

The Hierarchy of Forces

This perspective reveals why the fundamental forces seem so different in strength. It's not because they're actually different types of forces - it's because they operate in different substrate regimes:

Gravity operates when substrate displacement is minimal - massive objects create gentle, widespread pressure gradients. The forces are weak but long-ranging, like the barely perceptible currents in a vast ocean.

Electromagnetism involves moderate substrate displacement - moving charges create circulation patterns and pressure waves. The forces are stronger than gravity but still manageable, like the currents in a flowing river.

Nuclear forces emerge from extreme substrate displacement - multiple particles creating intense compression zones. The forces are enormous but short-ranged, like the explosive pressure when you squeeze water in a small container.

The Universal Constant

The equation contains one crucial parameter: the density of the substrate itself (ρ_QS). This single number determines the strength of every force in the universe. It's like the "stiffness" of space - how strongly the substrate resists being displaced.

Current calculations suggest this substrate density is incredibly small - about 1 $.0\times10^{-25}$ kg/m^3. That's almost nothing by everyday standards, atoms themselves have a density on the order of 100 kg/m^3 which explains why we've never directly detected the substrate. Yet this tiny density, when combined with the vast scales of space, creates all the forces that shape reality.

Think of it like this: even though air seems nearly weightless, atmospheric pressure can lift multi-ton aircraft and create devastating hurricanes. Similarly, even though the substrate is incredibly tenuous, its pressure gradients across cosmic distances create the forces that hold galaxies together and power stars.

From Falling Apples to Orbiting Galaxies

The elegance of this framework becomes clear when you realize it seamlessly connects phenomena across an almost incomprehensible range of scales:

A **coin falling** on Earth follows pressure gradients created by our planet's 1.32277 × 10^25 pounds.

The **Moon orbiting** Earth follows the same pressure gradients, just measured at a distance of 238,900 miles.

The **Earth orbiting** the Sun follows pressure gradients from our star's 4.385 × 10^30 pounds across 94.151 million miles.

Our **solar system orbiting** the galactic center follows pressure gradients from the Milky Way's central black hole across 26,000 light-years.

And **distant galaxies** moving apart follow pressure gradients created by the distribution of matter across billions of light-years.

Same equation. Same mechanism. Same substrate responding to matter at every scale imaginable.

The Mechanical Universe

What makes this framework revolutionary isn't just its mathematical elegance - it's the return to mechanical understanding. Instead of mysterious "fields" acting through empty space, we have pressure and flow in a medium. Instead of particles somehow "knowing" about distant forces, we have local interactions between matter and substrate.

Physics becomes intuitive again.

When you ask "How does gravity work?" The answer isn't "Spacetime curvature creates geodesic paths for matter to follow" (which explains nothing about the actual mechanism).

The answer is "Massive objects create internal substrate displacement zones with reduced pressure, while higher substrate pressure at the boundaries squeezes other objects toward these displacement regions with restoration forces that are seeking an equilibrium." (which makes perfect mechanical sense).

The Technological Promise

If this equation correctly describes reality, then the technological implications are staggering. Just as understanding electromagnetic equations led to electric motors, generators, radio, and computers, understanding substrate equations might lead to:

Gravity control: By manipulating substrate pressure gradients.

Advanced propulsion: Using substrate flows for movement.

Unlimited energy: Tapping into substrate pressure differentials.

Faster-than-light communication: Through substrate pressure waves.

These possibilities remain theoretical, but they follow naturally from mathematics. If forces are really pressure gradients in a substrate medium, then controlling those pressure gradients becomes an engineering problem rather than a physics impossibility.

The Test of Truth

Of course, an equation is only as good as its predictions. The true test of this framework isn't its elegance or intuitive appeal - it's whether it accurately describes what we observe in nature.

I put the substrate model to the test using data from the world's most powerful particle accelerator. The equation predicted that high-energy particle collisions should show characteristic 2.4 GeV oscillation signatures - specific energy intervals where the substrate's response to extreme particle density would create detectable patterns.

When I analyzed published ATLAS collision data with these predictions in mind, the results were encouraging. The predicted 2.4 GeV oscillations appeared to be present in the data, suggesting that the substrate model might indeed be detecting real physical effects rather than just providing mathematical descriptions.

The statistical significance of these results requires careful analysis, but the initial findings suggest that the substrate equation isn't just mathematically elegant - it might be describing actual mechanisms operating in high-energy physics experiments.

But we're still in the early stages of testing. The true validation will come as more experiments are designed specifically to look for substrate effects, and as the technological predictions begin to be realized.

The Journey Ahead

This equation might represent the beginning of a new chapter in physics - one where the universe becomes comprehensible again, where forces have clear mechanisms, and where technology can work with the fundamental structure of reality rather than against it.

In the chapters ahead, we'll explore how this simple equation explains each of the fundamental forces in detail, revealing the hidden mechanics behind everything from atomic structure to cosmic evolution. We'll see how the substrate model resolves long-standing mysteries in physics and opens pathways to technologies that seem impossible today.

But most importantly, we'll discover what it means to live in a mechanical universe - one where every force is understandable, every effect has a clear cause, and the deepest mysteries of physics become as intuitive as watching waves flow across a lake.

The substrate universe awaits. And it's more elegant than anyone imagined.

"The grand aim of all science is to cover the greatest number of empirical facts by logical deduction from the smallest number of hypotheses or axioms."

Einstein

CHAPTER 5

GRAVITY - THE GENTLE GIANT

WHY PLANETS DANCE TO PRESSURE'S TUNE

Stand up and jump. What brings you back down?

If you asked this question 400 years ago, most people would have said "things naturally fall toward the center of the Earth." If you asked 100 years ago, a physicist might have explained "massive objects warp spacetime, creating geodesic paths that matter follows."

But what if the real answer is simpler and more beautiful than either explanation? What if you fall back down because the Earth is constantly creating a gentle, invisible pressure system that extends far out into space - and you're swimming at the bottom of this pressure gradient like a fish in the deepest part of the ocean?

The Invisible Ocean Above Your Head

Right now, as you read this, you're sitting at the boundary of Earth's gravitational field - at the edge of the substrate displacement zone created by our planet. But instead of thinking about this as some mysterious force reaching out to hold you down, imagine it as a three-dimensional pressure system in the substrate that surrounds our planet.

Picture the Earth as a massive sphere displacing the substrate around it, creating a zone of slightly lower substrate through the mass and higher density that extends outward in all directions at its boundary.

This displacement creates pressure gradients - areas where the substrate pressure changes from point to point, like the pressure differences that create wind in our atmosphere.

The remarkable thing about these substrate pressure gradients is how they change with position. Here at Earth's surface, we experience the steepest pressure gradient because we're at the boundary between the displaced zone and the undisturbed substrate beyond. If we could somehow travel deep into the Earth, the substrate model predicts something fascinating: gravity would gradually weaken as we approached the center.

This happens because as we move deeper, we're traveling toward the center of the substrate displacement zone, where pressure gradients from all directions begin to balance out. At Earth's very center, the substrate displacement would be symmetrical in all directions, creating zero net pressure gradient and therefore zero gravitational effect - exactly what conventional physics predicts, but for entirely different mechanical reasons.

The remarkable thing about these substrate pressure gradients is how gentle they are. The pressure difference between your head and your feet - a distance of maybe six feet - is so tiny that sensitive instruments can barely measure it. Yet this minuscule pressure gradient is strong enough to keep you firmly planted on the ground and prevent you from drifting off into space.

It's like being at the bottom of the deepest ocean trench, where enormous pressures push on you from all directions. But unlike water pressure, substrate pressure at Earth's boundary squeezes you toward the planet's displacement zone. Just like hydrostatic pressure on a submarine hull, the higher substrate pressure surrounding Earth's cavitation zone pushes inward toward the center where substrate displacement is greatest.

From Coins to Galaxies

This pressure gradient explanation works identically whether you're dropping a coin or analyzing the motion of galaxies. Let's follow a pressure wave through different scales to see how the same mechanism operates everywhere:

The Falling Coin When you release a coin from your hand, it immediately begins responding to the substrate pressure gradient around Earth. The pressure difference between the coin's position and points slightly closer to Earth's center creates a net force pushing the coin downward.

The coin accelerates at 32.2 feet per second squared - not because Earth is mysteriously "attracting" it, but because this is exactly the acceleration you get when you divide the local substrate pressure gradient by the substrate's density. Every falling object experiences the same acceleration because they're all responding to the same pressure field.

The Orbiting Moon Scale up by a factor of about 60,000, and you reach the Moon's orbit. Here, the same Earth-created pressure gradient is still operating, but it's much weaker due to the distance. The Moon is constantly "falling" toward Earth just like your coin, but it's also moving sideways fast enough that it keeps missing our planet.

Think of the Moon as surfing along the pressure contours created by Earth's substrate displacement. It's like a marble rolling around the inner surface of a funnel - the marble wants to fall toward the center, but its sideways motion keeps it circling around the rim instead.

The Solar System Scale up another factor of 400, and you reach the Sun's influence. Now we're dealing with pressure gradients created by an object 333,000 times more massive than Earth. The Sun's substrate displacement is so enormous that it creates detectable pressure gradients across billions of miles.

All the planets are surfing these solar pressure contours. Mercury races around its tight orbit because it's deep in the Sun's pressure gradient, where the "downhill slope" is steep. Neptune crawls along its distant path because it's so far out that the pressure gradient is barely perceptible.

The Pressure Landscape

To really visualize how gravity works through substrate pressure, imagine you could see the pressure gradients around massive objects as a three-dimensional landscape.

Around Earth, you'd see a gentle bowl-shaped depression in the substrate pressure, deepest at Earth's center and gradually leveling out as you move away from the planet. The International Space Station orbits along a contour line in this pressure bowl, like a marble rolling around the rim of a shallow crater.

Around the Sun, you'd see a much deeper, wider depression. The planets orbit along different contour lines - Mercury following a tight circle near the bottom of the bowl, while distant Pluto traces a barely perceptible ellipse near the rim.

But here's where it gets truly spectacular: these pressure landscapes aren't separate. They're all part of one continuous, three-dimensional pressure map that extends throughout the entire solar system and beyond. Earth's pressure bowl sits inside the Sun's larger depression. The Moon's tiny pressure dimple sits inside Earth's bowl. It's pressure landscapes within pressure landscapes, all the way up to galactic scales.

The Tidal Mystery Solved

One of the most beautiful confirmations of substrate pressure mechanics comes from the tides. Twice each day, the oceans rise and fall in a rhythm that's puzzled humans for millennia. The standard explanation involves the Moon's gravity "pulling" on the oceans, but this has always been problematic - why do we get high tides on both the side of Earth facing the Moon and the side facing away?

The substrate model provides a crystal-clear answer. The Moon creates pressure gradients in the substrate around the Earth-Moon system. These pressure gradients don't "pull" on the oceans - they create a complete three-dimensional pressure field that pushes and squeezes the oceans into an elliptical shape.

The Moon's displacement zone creates regions where substrate pressure is reduced. The higher substrate pressure in surrounding areas squeezes the oceans toward these displacement zones, creating the characteristic elliptical tidal bulges. The side of Earth farthest from the Moon also experiences lower substrate pressure (it's farther from the Moon's pull), so the oceans get pushed outward there too.

The result is that Earth's oceans get squeezed into an elliptical shape, with high tides at both ends of the ellipse and low tides on the sides. As Earth rotates through this elliptical pressure field, any given location experiences two high tides and two low tides each day.

It's not mysterious action at a distance - it's pressure mechanics, as understandable as squeezing a water balloon.

The Galactic Dance

The same pressure gradient mechanism that explains falling coins and orbiting moons also explains the motion of our entire galaxy. The Milky Way contains over 100 billion stars, all orbiting around a common center in a cosmic dance that takes 230 million years to complete.

At the galaxy's heart sits a supermassive black hole - an object so dense that it creates substrate displacement unlike anything else in the universe. This creates pressure gradients that extend across the entire galaxy, organizing the motion of billions of stars into elegant spiral patterns.

Our solar system, located about 26,000 light-years from the galactic center, orbits along one of these pressure contours at a speed of 514,000 miles per hour. We're surfing a pressure gradient created by a black hole millions of times more massive than our Sun, yet the mechanism is identical to a coin falling in your hand.

Why Gravity is Universal

One of the most remarkable features of gravity is its universality. Unlike electromagnetic forces, which only affect charged objects, or nuclear forces, which only operate at atomic scales, gravity affects everything equally. A feather and a hammer dropped on the Moon fall at exactly the same rate.

The substrate model explains this beautifully. Every object - regardless of its composition, charge, or internal structure - displaces the substrate simply by existing. The substrate doesn't "care" whether an object is made of lead or feathers, whether it's positively charged or neutral, whether it's a single atom or a massive star. It responds to the presence of matter itself.

This is why Einstein's principle of equivalence works so perfectly: you can't tell the difference between being accelerated upward by a rocket and being pulled downward by gravity, because both involve the same substrate pressure mechanics. In both cases, you're being pushed by pressure gradients in the substrate medium.

The Porous Submarine: Why Gravity Persists Deep Underground

Imagine you could dig a hole straight down through the Earth - not just the 7.6 miles of the Kola Superdeep Borehole, but a full hundred miles into the planet's interior. Would gravity disappear as you moved away from Earth's surface? Would a titanium box at that depth float weightlessly in some kind of gravitational void?

The substrate model reveals why this intuition is completely wrong. Earth doesn't act like a solid submarine that completely displaces water from its interior. Instead, it behaves like a **porous submarine** - one where the "water" (substrate) is constantly being squeezed out but continuously trying to flow back in.

When Earth's massive bulk displaces substrate, it doesn't create an empty void at the planet's center. Rather, it establishes a **dynamic pressure gradient** where substrate density decreases toward the center but never reaches zero. Think of squeezing a sponge underwater - you can compress the water out, but it never becomes completely dry, and the water constantly pushes back trying to restore its original state.

This is why gravity remains strong even deep underground. Your hypothetical titanium box 100 miles down still sits within Earth's vast **substrate displacement boundary** - the region where the planet's total mass has squeezed substrate density below normal levels. The pressure gradients that create gravitational acceleration don't disappear with depth; they remain because the box is still within the zone where substrate is actively being displaced by Earth's accumulated mass.

The restoration forces are constantly at work, with substrate trying to flow back toward equilibrium density. But Earth's enormous mass continuously maintains the displacement, creating ongoing pressure dynamics rather than static empty space. Gravity persists because the substrate "knows" it's been squeezed away from its preferred density and maintains pressure gradients pointing toward the displacement center.

This porous submarine model explains why astronauts in the International Space Station still experience about 90% of Earth's surface gravity - they're still well within the substrate displacement boundary, even at 250 miles above the surface. The displacement zone extends far beyond the planet itself, creating gravitational effects throughout the region where Earth's mass has altered substrate density.

The Weakness and the Strength

Gravity is by far the weakest of the fundamental forces. The electromagnetic force between two electrons is about 10^40 times stronger than the gravitational force between them. Yet gravity dominates the large-scale structure of the universe. Why?

The substrate model reveals the answer. Gravity operates through very gentle pressure gradients that extend across enormous distances. Each individual atom contributes a tiny amount to the overall pressure field, but when you have 10^50 atoms working together (as in a planet or star), the cumulative effect becomes enormous.

It's like the difference between a gentle breeze and a hurricane. A single puff of air has almost no effect, but when trillions of air molecules move together in the same direction, they can level buildings and uproot trees. Similarly, individual atoms create negligible substrate displacement, but when countless atoms work together, they shape the motion of entire solar systems.

The Technology of Tomorrow

If gravity really operates through substrate pressure gradients, then the technological implications are staggering. Just as we learned to generate and control electromagnetic fields once we understood their mechanism, we might learn to manipulate gravitational fields once we master substrate engineering.

Imagine being able to create artificial pressure gradients in the substrate around a spacecraft. By generating substrate displacement zones in the direction you want to travel, you could create pressure differentials that squeeze your ship toward these zones through boundary pressure effects. You'd be surfing substrate pressure waves like a cosmic sailboat.

Or consider the possibility of gravitational shielding. If you could somehow increase the substrate density in a localized area, you might be able to block or redirect gravitational pressure gradients. A room with gravitational shielding would become a zero-gravity environment, no matter where it was located.

These technologies remain purely theoretical, but they follow logically from the substrate model. If gravity is pressure, then controlling pressure becomes an engineering challenge rather than a physics impossibility.

The Elegant Truth

Perhaps the most beautiful aspect of understanding gravity through substrate mechanics is how it transforms our perception of the cosmos. Instead of objects mysteriously pulling on each other across empty space, we have a universe filled with flowing pressure gradients, creating elegant patterns of motion from the smallest scales to the largest.

You're not held to Earth by some mysterious force - you're swimming in a pressure gradient so gentle and constant that you've never known anything else. The Moon doesn't orbit Earth because of magical attraction - it's surfing pressure contours in the substrate medium. The galaxies don't hold together through dark matter - they're organized by pressure landscapes created by supermassive black holes.

It's a mechanical universe, operating through principles as understandable as water flowing downhill or air moving from high pressure to low pressure. The only difference is that it's happening in a medium so subtle that we're just beginning to detect its effects.

Gravity isn't mysterious anymore. It's pressure mechanics on a cosmic scale, and it's more elegant than anyone ever imagined.

"I have not yet been able to discover the cause of these properties of gravity from phenomena, and I frame no hypotheses."

Newton

CHAPTER 6

ELECTROMAGNETISM - NATURE'S WHIRLPOOLS

WHEN THE SUBSTRATE STARTS SPINNING

Take a strong magnet and sprinkle iron filings around it. Watch as they arrange themselves into beautiful curved patterns, revealing invisible "field lines" that seem to emerge from one end of the magnet and curve around to the other end.

What are you actually seeing? Most physics textbooks will tell you these are abstract mathematical representations of magnetic field strength and direction. But what if they're showing you something far more concrete - the actual circulation patterns of the substrate itself?

What if every magnet is creating whirlpools in the invisible medium that fills space, and those iron filings are simply revealing the flow?

The Spinning Substrate

While gravity operates through pressure gradients in the substrate - regions where the substrate density changes from place to place - electromagnetism involves something more dynamic: circulation. The substrate can not only be compressed and stretched, but it can also rotate and swirl, creating patterns that persist across space and time.

But here's a crucial difference: unlike substrate displacement effects (gravity and inertia) which affect all matter equally, substrate circulation patterns only couple strongly with matter that has its own internal circulation - primarily systems with mobile charged particles.

This explains why you don't feel magnetic fields the way you feel gravitational effects. Your body is mostly composed of neutral atoms with balanced, stable charge distributions. While magnetic fields create circulation patterns in the substrate all around you, your tissues don't have the internal circulation patterns needed to couple with these substrate flows.

Metals, however, are different. They contain mobile electrons that can move freely through the material, creating internal circulation patterns that can couple with substrate circulation. When substrate circulation passes through a metal, it can drag these mobile electrons along, creating electric currents. When the substrate circulation changes, it can induce forces on the metal by interacting with its internal electron circulation patterns.

The Frequency Tuning Effect

Think of electromagnetic coupling like tuning a radio station. Substrate circulation patterns are constantly flowing through space at various frequencies and characteristics - like radio waves broadcasting on different stations throughout the electromagnetic spectrum.

Materials respond based on their internal "tuning":

Metals (Perfect Reception): Mobile electrons can adjust their circulation patterns to match passing substrate circulation. When substrate circulation at the "right frequency" flows through, the electrons can tune into it. Once matched, the substrate circulation essentially "grabs the electrons by the hand" and drags the whole chain along, creating electric current.

Biological Tissues (Wrong Frequency): Your cells contain mostly fixed charge patterns around stable molecules. They can't adjust to match the passing substrate circulation, so electromagnetic fields pass through without coupling - like radio waves passing through a radio tuned to the wrong station.

Magnetic Materials (Resonant Reception): Iron and other magnetic materials have internal magnetic domains (electron spin patterns) that can align with specific substrate circulation patterns. They act like perfectly tuned antennas that couple strongly with magnetic substrate circulation.

This frequency-matching explains why iron filings align with magnetic fields while wood chips don't, why metal detectors work on coins but not plastic, and why only certain materials become magnetic. The substrate circulation is present everywhere, but only materials with matching internal patterns can "tune in" to couple with it strongly enough to experience noticeable effects.

Electric Charge: The Persistent Whirlpool

What is electric charge? The standard answer is that it's a fundamental property of certain particles, but this doesn't really explain what it *is* - it just gives it a name.

The substrate model suggests a much more concrete answer: electric charge represents persistent circulation patterns in the substrate. A positively charged particle is surrounded by the substrate rotating in one direction, while a negatively charged particle is surrounded by substrate rotating in the opposite direction.

Think of a charged particle as being like a tiny motor that's constantly stirring the substrate around it. Once this stirring pattern is established, it tends to persist - the substrate keeps circulating in the same direction, creating a stable, long-lasting whirlpool that extends outward through space.

This immediately explains several mysteries about electric charge:

Why charge is conserved: You can't create or destroy circulation patterns from nothing. If you want to create a clockwise whirlpool, you inevitably create a counterclockwise whirlpool somewhere else to balance it out. Positive and negative charges always appear in equal amounts because circulation patterns must balance.

Why like charges repel: Two whirlpools spinning in the same direction interfere with each other. Their circulation patterns conflict, creating pressure zones that push the particles apart.

Why opposite charges attract: Whirlpools spinning in opposite directions can mesh together like gears, creating flow patterns that pull the particles toward each other.

Why charge comes in discrete amounts: Stable circulation patterns in the substrate can only exist in certain configurations, just like stable whirlpools in water can only form with certain amounts of circulation.

Magnetic Fields: Whirlpools Made Visible

When you see those iron filings arranging themselves around a magnet, you're witnessing something remarkable: you're actually seeing the circulation patterns in the substrate through materials that can couple with them.

A magnetic field is simply a region where the substrate is circulating in an organized way. The iron filings act like tiny detectors, aligning themselves with the direction of substrate circulation because iron contains internal magnetic domains - regions where electron spins create small circulation patterns that can couple with the substrate circulation.

Most materials don't respond this way. Sprinkle wood chips or plastic particles around a magnet, and they remain randomly oriented because these materials lack the internal circulation patterns needed to couple with substrate flows. But materials with mobile electrons (like metals) or aligned magnetic domains can interact with substrate circulation patterns, making the invisible flows visible.

This explains why magnetic fields have the properties they do:

Magnetic field lines form closed loops: Just like whirlpools in water, circulation patterns in the substrate must close back on themselves. The substrate that flows out of one side of a magnet must flow back into the other side.

Magnetic fields have direction: The circulation has a definite direction - clockwise or counterclockwise when viewed from different angles. This is what we measure as the magnetic field direction, revealed by materials that can couple with the substrate circulation.

You can't have isolated magnetic poles: You can't have a whirlpool that only spins out but never spins back in. Every magnetic north pole must be paired with a south pole, because circulation patterns must be complete.

Why you don't feel magnetic fields: Unlike gravitational effects that couple with all matter through substrate displacement, magnetic effects only couple strongly with materials that have internal circulation patterns to match the substrate circulation.

The Dance of Moving Charges

One of the most beautiful demonstrations of substrate mechanics comes from watching what happens when electric charges move. When you move a charged particle - essentially dragging a circulation pattern through the substrate - you create additional circulation perpendicular to the motion.

It's like dragging a spinning ball through honey. The ball's rotation creates swirling patterns in the honey that extend outward from the ball's path. Similarly, when a charged particle moves through the substrate, its existing circulation pattern gets twisted and stretched, creating new circulation patterns perpendicular to its motion.

This is why moving electric charges create magnetic fields. The particle's charge (its persistent circulation pattern) combines with its motion to create new, perpendicular circulation patterns that we detect as magnetism.

This also explains why electric currents in wires create magnetic fields around them. Millions of electrons moving through the wire are each dragging their circulation patterns along, creating a combined circulation system that extends outward from the wire in concentric circles.

Electromagnetic Induction: Changing the Station

Now comes one of the most elegant demonstrations of substrate mechanics: electromagnetic induction. When you change a magnetic field near a conducting wire, you automatically create electric forces in that wire. This is how electric generators, transformers, and induction motors work.

In the substrate model, this makes perfect mechanical sense through the radio tuning analogy. When you change the circulation patterns in one region (changing the magnetic field), you're essentially "changing the station" - shifting the frequency characteristics of the substrate circulation. This creates new circulation patterns that can suddenly couple with materials that weren't responsive before.

It's like doing a frequency sweep across a radio dial. Materials that couldn't tune into the original substrate circulation might suddenly find themselves able to couple with the new circulation patterns. When these new patterns reach charges in a nearby wire, they can suddenly "grab onto" electrons that weren't being affected before, creating electric forces and currents.

This explains why:

Moving a magnet near a wire generates electricity: The changing circulation patterns create frequency sweeps that can couple with electrons in the wire.

Changing electric currents create changing magnetic fields: Moving charges create new circulation patterns with different frequency characteristics.

Electromagnetic waves can propagate through space: They're substrate circulation patterns sweeping through frequency ranges as they propagate.

Radio Technology Through Substrate Tuning: Every electromagnetic device becomes a substrate circulation tuner.

Antennas: Precisely designed structures that couple optimally with specific substrate circulation frequencies.

Radio receivers: Circuits tuned to detect and amplify specific circulation patterns from the substrate.

Transmitters: Devices that create controlled substrate circulation patterns at chosen frequencies.

Transformers: Devices that couple substrate circulation between different circuits through frequency matching.

Light: Pressure Waves in the Substrate Ocean

This brings us to one of the most profound implications of substrate mechanics: light isn't a mysterious "self-propagating electromagnetic disturbance in empty space." It's pressure waves in the substrate medium, exactly analogous to sound waves in air or water waves in the ocean.

When electrons vibrate in the antenna of a radio transmitter, they create oscillating circulation patterns in the substrate. These changing circulation patterns generate pressure waves that propagate outward at the speed of light - which is simply the speed of pressure waves in the substrate medium.

The speed of light is determined by the substrate's mechanical properties - its density and elasticity - exactly the way the speed of sound is determined by air's density and compressibility.

This explains several puzzling features of light:

Why light has a constant speed: Just like sound waves in air always travel at the same speed regardless of how they're generated, pressure waves in the substrate always propagate at the same speed.

Why light can have different frequencies: Just like you can create high-pitched or low-pitched sound waves, you can create high-frequency or low-frequency pressure waves in the substrate. We perceive these as different colors.

Why light exhibits wave-particle duality: Sound waves sometimes behave like waves (interference, diffraction) and sometimes like particles (discrete energy packets). Substrate pressure waves behave the same way.

Why light doesn't need a medium to propagate: Light *does* need a medium - the substrate. We just haven't recognized it as such because the substrate is so perfectly transparent to matter.

From Radios to MRI Machines

Understanding electromagnetism as substrate circulation patterns transforms how we see electromagnetic technology. Every device that uses electricity or magnetism is essentially a substrate circulation manipulator:

Electric motors work by creating circulation patterns in the substrate (magnetic fields) that interact with the circulation patterns around moving charges (electric currents) to produce mechanical force.

Generators work by mechanically disrupting substrate circulation patterns (moving conductors through magnetic fields) to create pressure waves that push on charges, generating electricity.

Transformers work by using changing circulation patterns in one coil to create pressure waves that generate new circulation patterns in another coil.

Radio antennas work by creating oscillating circulation patterns that generate pressure waves (radio waves) that propagate through the substrate to distant receivers.

MRI machines work by using powerful circulation patterns to align the tiny circulation patterns associated with atomic nuclei, then detecting the pressure waves these nuclei emit when they return to their normal orientations.

Every electromagnetic device is manipulating the same substrate medium through circulation and pressure wave mechanics.

The Electromagnetic Spectrum: A Symphony of Substrate Frequencies

What we call the electromagnetic spectrum - from radio waves to gamma rays - is simply different frequencies of substrate circulation patterns, like having access to a cosmic radio dial with an enormous range of stations.

Radio waves are low-frequency, long-wavelength substrate circulation patterns - like the slow, rolling circulation in a lazy river.

Microwaves are faster circulation patterns that happen to match the natural resonance frequencies of water molecules, which is why microwave ovens work - they broadcast on the frequency that water molecules can "tune into."

Infrared radiation consists of circulation patterns that we feel as heat - they're at frequencies that efficiently couple with the vibrational patterns of most molecules.

Visible light covers the narrow range of circulation frequencies that our eyes evolved to detect - probably because these frequencies penetrate our atmosphere most efficiently and provide the best information about our environment.

Ultraviolet, X-rays, and gamma rays are extremely rapid circulation patterns in the substrate, carrying so much energy that they can disrupt the circulation patterns that hold atoms and molecules together.

All of these are the same phenomenon - substrate circulation patterns at different frequencies - just like having access to AM radio, FM radio, television, and wireless internet all operating through the same EM medium but at different frequency bands.

Materials respond selectively: Just as your radio can be tuned to receive FM but not AM, different materials can couple with different substrate circulation frequencies:

Radio antennas: Designed to couple perfectly with specific circulation frequencies

Solar panels: Tuned to couple with visible light circulation patterns

X-ray detectors: Made from materials that can couple with high-frequency substrate circulation

Biological molecules: Often have natural resonance frequencies that couple with specific electromagnetic frequencies

The Technological Revolution Ahead

If electromagnetism really operates through substrate circulation patterns, then we're only beginning to explore the technological possibilities. Just as understanding air pressure led to flight and understanding water pressure led to hydraulics, understanding substrate pressure and circulation might lead to technologies that seem impossible today.

Imagine being able to create custom circulation patterns in the substrate - artificial magnetic fields with any shape or strength you desire. You could create:

Magnetic bottles that could contain fusion reactions with perfect efficiency, solving the world's energy problems.

Circulation shields that could protect spacecraft from radiation by deflecting charged particles with precisely controlled substrate whirlpools.

Substrate antennas that could generate electromagnetic waves with perfect efficiency, creating communication systems that work across interstellar distances with minimal power.

Circulation computers that could process information using substrate flow patterns, potentially achieving computational speeds that dwarf anything possible with electronic circuits.

These possibilities remain theoretical, but they follow naturally from understanding electromagnetism as substrate mechanics rather than mysterious field interactions.

The Beauty of Circulation

Perhaps the most satisfying aspect of understanding electromagnetism through substrate mechanics is how it makes the phenomenon intuitively beautiful. Instead of abstract mathematical fields existing in empty space, we have elegant circulation patterns in a dynamic medium - whirlpools and currents and pressure waves all interacting in perfectly mechanical ways.

Everytime you flip a light switch, you complete circulation patterns in the substrate that travel at light speed to generate pressure waves that push on electrons in the light bulb's filament.

Everytime you use a compass, you're detecting the Earth's massive circulation patterns in the substrate.

Everytime you listen to the radio, you're receiving pressure waves in the substrate that were generated by circulation patterns hundreds of miles away.

The electromagnetic world isn't mysterious anymore. It's a symphony of circulation and pressure in the invisible ocean that surrounds us - and it's more elegant than anyone ever imagined.

"The nature of electric current is still very obscure, and all we can do is to use mechanical illustrations which, though they may not be true, enable us to picture to ourselves what is going on."

Maxwell

CHAPTER 7

THE NUCLEAR MYSTERIES

WHEN PARTICLES CREATE COSMIC WHIRLPOOLS

Deep inside every atom in your body lies a mystery that has puzzled physicists for nearly a century. At the heart of each atom sits a tiny nucleus - so small that if an atom were the size of a football stadium, the nucleus would be smaller than a marble sitting on the fifty-yard line. Yet this microscopic speck somehow holds together with forces so powerful that they dwarf every other force in nature.

The energy locked in atomic nuclei powers the Sun, lights up entire cities through nuclear reactors, and in its most destructive form, can level mountains. Where does this tremendous energy come from? And why does it only operate across distances smaller than the nucleus itself?

The substrate model provides an answer that's both elegant and startling: nuclear forces aren't separate forces at all. They're what happens when particles become so tightly packed that they create extreme displacement zones in the substrate - cosmic whirlpools so intense that they make electromagnetic forces look like gentle breezes.

The Nuclear Void

To understand nuclear forces, we need to imagine something almost impossible to visualize: what happens when you try to squeeze something the size of several bowling balls into a space smaller than a marble?

In the substrate model, every particle displaces some amount of the surrounding substrate medium. For individual particles spread out in space, this displacement is minimal - like dropping a pebble into a lake. But when multiple particles cluster together in the incredibly tiny volume of an atomic nucleus, their combined displacement effects become extreme.

Picture trying to push several beach balls underwater simultaneously in a small swimming pool. Each ball wants to displace its own volume of water, but there simply isn't enough room. The water around them gets compressed into an incredibly small space, creating enormous pressure. The water pushes back with tremendous force, trying to equalize the pressure and prevent the beach balls from occupying the same space.

This is essentially what happens in atomic nuclei, but with substrate instead of water. Multiple protons and neutrons clustered in a volume millions of times smaller than the atom itself create what we might call a "nuclear void" - a region where the substrate has been displaced so dramatically that it creates pressure effects unlike anything else in nature.

The Strong Force: Extreme Pressure Response

The nuclear strong force - the most powerful force in nature - is the substrate's response to this extreme displacement. When particles are packed so tightly that they create substantial substrate voids, the surrounding substrate pushes back with enormous pressure, binding the nucleus together.

If you could somehow create a perfect vacuum bubble inside a block of steel, the steel would collapse inward with tremendous force to fill that vacuum. Similarly, when nuclear particles create substrate voids, the surrounding substrate pushes inward with enormous force to fill those voids.

This explains several mysterious features of the strong nuclear force:

Why it's so powerful: The pressure created by extreme substrate displacement is enormous - thousands of times stronger than electromagnetic forces. It's like comparing the gentle pressure of a breeze to the crushing pressure at the bottom of the ocean.

Why it's so short-ranged: The extreme pressure effects only extend across the size of the substrate void itself - roughly the diameter of the nucleus. Beyond this distance, the substrate displacement becomes manageable, and the pressure drops off rapidly.

Why it affects all nucleons equally: The strong force doesn't care whether a particle is a proton or neutron, just as water pressure doesn't care what material your submarine is made of. All that matters is how much substrate each particle displaces.

Why it has a repulsive core: When particles get too close together, their individual displacement zones start to overlap, creating even more extreme substrate displacement. The substrate pushes back even harder, preventing the particles from occupying the same space.

Nuclear Binding Energy: The Substrate Pressure Well

The energy that holds atomic nuclei together - nuclear binding energy - comes from the substrate pressure well created by the nuclear void. Just as it takes energy to lift an object out of a gravitational well, it takes enormous energy to pull particles out of a nuclear substrate pressure well.

When you split a heavy nucleus (nuclear fission) or fuse light nuclei together (nuclear fusion), you're essentially rearranging these substrate pressure wells. Some arrangements create deeper wells than others. When you move from a shallow well to a deeper one, the excess energy gets released - and because the substrate pressure effects are so intense, the amount of energy released is enormous.

This is why nuclear reactions release millions of times more energy than chemical reactions. Chemical reactions involve rearranging the relatively gentle electromagnetic circulation patterns between molecules. Nuclear reactions involve rearranging the extreme substrate pressure systems inside atomic nuclei.

The Weak Force: Substrate Instability

The weak nuclear force - responsible for certain types of radioactive decay - represents a different aspect of substrate mechanics: instability in the substrate pressure systems.

Some nuclear configurations create substrate pressure systems that aren't quite stable. The substrate displacement is so extreme, and the pressure gradients so intense, that the system occasionally undergoes spontaneous rearrangements to reach a more stable configuration.

Imagine a house of cards built in a windstorm. Most of the time, the cards hold their positions despite the buffeting wind. But occasionally, a particularly strong gust causes the structure to collapse and rearrange into a different, more stable configuration.

This is what happens in weak nuclear decay. The substrate pressure system around certain nuclei is metastable - it can persist for a while, but it's not the most stable possible configuration. Eventually, quantum fluctuations in the substrate cause the system to spontaneously rearrange, ejecting particles and energy in the process.

Atomic Structure: Electrons in Pressure Wells

This understanding of nuclear forces as extreme substrate displacement revolutionizes how we think about atomic structure. Instead of electrons somehow "orbiting" the nucleus like tiny planets, we can picture them as being trapped in the pressure wells created by the nuclear void.

The nuclear substrate displacement doesn't just affect other nuclear particles - it creates pressure gradients that extend outward from the nucleus, getting weaker with distance but still detectable. Electrons, being much lighter than nuclear particles, get caught in these pressure gradients like marbles rolling around the rim of a bowl.

The "electron orbitals" that chemistry students memorize aren't mysterious quantum mechanical probability clouds - they're the three-dimensional shapes of the substrate pressure wells around different types of nuclei. Electrons naturally settle into the areas where the substrate pressure gradients create stable equilibrium points.

Chemical Bonding: Overlapping Pressure Wells

Chemical bonding - the forces that hold molecules together - emerges naturally when the pressure wells of different atoms overlap and interact.

When two atoms approach each other, their nuclear substrate displacement zones begin to interact. In some cases, the overlapping pressure systems create new, combined pressure wells that are more stable than the separate atomic wells. Electrons from both atoms can then settle into these combined pressure systems, effectively "sharing" the electrons between nuclei and creating chemical bonds.

Different types of chemical bonds correspond to different ways the nuclear pressure systems can combine:

Covalent bonds form when nuclear pressure wells overlap symmetrically, creating shared electron trapping regions between the nuclei.

Ionic bonds form when one nucleus creates such intense pressure gradients that it captures electrons from another nucleus, creating charged ions that attract each other.

Metallic bonds form when many nuclear pressure wells overlap in an extended network, creating electron trapping regions that extend throughout the material.

The Hierarchy Explained

The substrate model finally explains why the fundamental forces have such different strengths.

It's not because they're fundamentally different types of forces - it's because they operate in different substrate displacement regimes across an enormous spectrum with sharp boundary transitions:

Gravity ($\varphi \approx 1$) operates with minimal substrate displacement around particle densities of ~$10^{29}/m^3$. The effects are gentle and long-ranged, like ripples on a calm lake. The substrate maintains nearly its background density with only tiny perturbations.

Electromagnetism ($\varphi \approx 10^{-6}$) involves moderate substrate displacement at particle densities around $10^{35}/m^3$. Here the substrate develops organized circulation patterns and quantum resonance effects, creating forces stronger than gravity but still manageable.

Nuclear forces ($\varphi \approx 10^{-35}$) emerge from extreme substrate displacement at particle densities of ~$10^{44}/m^3$. Multiple nucleons create such intense compression that the substrate approaches complete void formation. The forces are enormous but short-ranged, like the explosive pressure when the substrate approaches its breaking point.

Black holes ($\varphi \rightarrow 0$) represent the absolute extreme at critical particle densities approaching $10^{50}/m^3$. Here the substrate displacement becomes complete - the medium fails entirely, creating genuine voids where no substrate exists at all.

The Piecewise Structure of Reality

What makes this hierarchy particularly important is that substrate physics operates through **sharp boundary transitions** rather than smooth changes. The substrate doesn't gradually shift from one regime to another - it undergoes dramatic phase changes at specific displacement thresholds.

We can think of the piecewise structure of the nuclear force like the behavior of water as it transitions to steam. At lower temperatures, water behaves as a liquid — cohesive, flowing, and relatively dense. But as energy is added and the boiling point is reached, the structure of water undergoes a sharp transition. The same molecules suddenly behave very differently, dispersing as gas.

This change isn't gradual — it's abrupt and defined by a boundary condition: the boiling point. In the same way, the nuclear force operates under different structural rules depending on energy, distance, or the internal state of nucleons. Each "region" is governed by a different rule or coupling behavior — a piece of the overall function — and the transitions between them are sharp, like phase changes.

Nuclear Regime Boundary ($\varphi \approx 10^{-35}$):

Above this threshold: Strong force dominates through extreme substrate compression effects.

Below this threshold: Weak force effects emerge as the substrate enters instability and oscillation modes.

The sharp transition explains why nuclear physics seems to have two completely different force types rather than a smooth spectrum.

Event Horizon Boundary ($\varphi \to 0$):

Above this threshold: Extreme substrate displacement creates intense gravitational effects, but the medium still exists.

At this exact threshold: Complete substrate failure creates the event horizon boundary.

Below this threshold: No substrate medium exists (the singularity region).

The sharp transition explains why black holes have such definite, measurable event horizon boundaries.

Why Boundaries Create Bizarre Behavior: The most mysterious phenomena in physics occur precisely at these substrate regime boundaries. This isn't coincidence - it's where the substrate medium undergoes dramatic phase changes that create effects unlike anything in the normal regimes:

Strong/weak nuclear transitions: Particle decay and transformation effects at the nuclear boundary.

Event horizon effects: Time dilation and causality limits at the gravitational boundary.

Quantum transitions: Wave-particle behavior changes at quantum field boundaries.

The famous hierarchy of force strengths - gravity weakest, electromagnetic intermediate, nuclear strongest - is simply the hierarchy of substrate displacement intensities operating through a piecewise structure. But even nuclear forces, extreme as they are, represent only the penultimate step before complete substrate failure at black hole boundaries.

Radioactivity: When Pressure Systems Fail

Radioactive decay occurs when nuclear substrate pressure systems become so extreme that they can't maintain stability. There are several ways this can happen:

Alpha decay occurs when a nucleus is so large that its substrate displacement zone becomes unstable. The system spontaneously ejects a cluster of particles (an alpha particle) to reduce the displacement and achieve a more stable configuration.

Beta decay occurs when the internal substrate pressure system favors a different balance of protons and neutrons. The system spontaneously converts one type of particle into another, ejecting the excess energy and particles.

Gamma decay occurs when a nucleus is in an excited substrate pressure configuration. The system relaxes to a lower-energy configuration, emitting the excess energy as high-frequency pressure waves (gamma rays) in the substrate.

Each type of radioactive decay represents the substrate pressure system finding a path to greater stability.

The Nuclear Technology Revolution

Understanding nuclear forces as substrate pressure effects opens up revolutionary technological possibilities. Instead of crude approaches like smashing nuclei together or waiting for natural decay, we might learn to manipulate substrate pressure systems directly.

Imagine being able to create artificial substrate displacement zones that could:

Control fusion reactions by creating precisely shaped substrate pressure wells that guide nuclei together under optimal conditions, making fusion power practical and safe.

Neutralize radioactive waste by manipulating the substrate pressure systems around unstable nuclei, forcing them into stable configurations.

Create new elements with desired properties by engineering specific substrate displacement patterns that don't occur naturally.

Generate nuclear energy on demand by controlling the substrate pressure well depths and triggering nuclear reactions precisely when and where needed.

These technologies would require understanding substrate mechanics at a level we haven't achieved yet, but they follow naturally from the pressure well model of nuclear forces.

The Plausibility

Perhaps the most beautiful aspect of understanding nuclear forces through substrate mechanics is how it transforms the atomic world from mysterious to mechanical. Instead of bizarre quantum effects and incomprehensible forces, we have extreme pressure systems operating according to principles as understandable as fluid dynamics.

Every atom in your body is a tiny pressure system in the substrate medium. The carbon in your muscles, the calcium in your bones, the iron in your blood - each element represents a different configuration of substrate displacement and pressure well geometry. Chemical reactions are substrate pressure systems rearranging into more stable configurations. Nuclear reactions are extreme substrate pressure systems undergoing dramatic rearrangements.

You're not made of mysterious quantum particles following incomprehensible laws. You're made of substrate pressure patterns - elegant, mechanical, and perfectly understandable once you recognize the medium they're operating in and the piecewise structure that governs their behavior.

The nuclear mysteries aren't mysterious anymore. They're pressure mechanics on an almost unimaginably intense scale, operating through sharp boundary transitions.

"What we observe is not nature itself, but nature exposed to our method of questioning."

Heisenberg

CHAPTER 8

THE VIBRATIONAL BLUEPRINT

HOW SUBSTRATE PATTERNS HOLD EVERYTHING TOGETHER

Take a handful of sand and sprinkle it evenly across a metal plate. Attach the plate to a speaker and play a single, pure musical tone. Watch as something almost magical happens: the randomly scattered sand begins to move, flowing into intricate geometric patterns. Lines, circles, stars, and complex symmetrical designs emerge from what was chaos just moments before.

Change the frequency, and the pattern instantly transforms into something completely different. Each frequency creates its own unique, stable pattern. Turn off the sound, and the sand returns to random scatter.

This phenomenon, called cymatics, reveals something profound about the relationship between vibration and structure. Sound waves - pressure oscillations traveling through air - can create and maintain stable patterns in physical matter. But what if this principle operates far beyond what we can see with sand and speakers?

What if every molecule in your body, every crystal in a rock, every complex protein that makes life possible, is held together by invisible vibrational patterns in the substrate medium? What if the entire physical world is one vast cymatics demonstration, with substrate oscillations creating the stable patterns we call atoms, molecules, and living structures?

The Invisible Orchestra

We've seen how the substrate medium responds to matter through pressure gradients (gravity) and circulation patterns (electromagnetism). But the substrate might be capable of something even more elegant: resonance patterns that create and maintain the complex structures of matter itself.

The beautiful thing is that these resonance patterns are powered by the same mechanism that creates gravity - the substrate's constant tendency to restore equilibrium pressure.

When matter displaces substrate, it doesn't just create static pressure differences. The substrate continuously tries to restore its natural density, but like a cork bobbing on water after being pushed down, this restoration process creates oscillations. The substrate repeatedly overshoots equilibrium, creating rhythmic pressure waves around every piece of matter.

These pressure oscillations become the driving force for substrate resonance patterns. The same restorative pressure that creates gravitational effects also provides the energy that maintains molecular structure. It's a self-sustaining system - matter creates its own structural maintenance energy simply by existing in the substrate medium.

Imagine the substrate as being capable of oscillation, like a vast, invisible drumhead that can vibrate at countless different frequencies simultaneously. Each piece of matter creates its own oscillation pattern through substrate displacement, and these patterns can interact, interfere, and combine to create incredibly complex three-dimensional structures.

Fundamental particles aren't solid objects but rather stable resonance patterns in this oscillating substrate - persistent vibrational structures powered by the substrate's endless attempts to restore equilibrium around displaced regions.

This transforms our understanding of what matter actually is. Instead of tiny, solid building blocks assembled into larger structures, we might have hierarchical resonance patterns in the substrate medium - oscillations creating particles, particles creating atoms, atoms creating molecules, molecules creating the complex structures of life.

Molecular Architecture Through Resonance

Consider a simple water molecule - two hydrogen atoms bonded to one oxygen atom in a precise angular arrangement. Chemistry textbooks explain this through "electron sharing" and "orbital overlap," but these explanations never quite answer the fundamental question: what actually holds the atoms in exactly that geometric arrangement, and where does the energy come from to maintain it?

The substrate resonance model suggests a beautifully mechanical answer: the water molecule is a standing wave pattern in the substrate created by the oscillatory restorative pressure around the hydrogen and oxygen nuclei.

Here's the elegant mechanism: Each nucleus displaces substrate, creating pressure imbalances. The substrate continuously tries to restore equilibrium around each nucleus, creating oscillatory pressure waves. When multiple nuclei are close together, their individual oscillation patterns interfere with each other, creating complex standing wave interference patterns.

The 104.5-degree bond angle of water isn't arbitrary - it's the geometric arrangement where the substrate oscillation patterns of two hydrogen nuclei and one oxygen nucleus create the most stable constructive interference. The molecules don't need external energy to maintain this structure because the energy comes from the substrate's endless attempts to restore equilibrium around the nuclear masses.

Picture three tuning forks arranged in a triangle, but instead of needing to strike them repeatedly, they're powered by springs that keep them oscillating. If the springs are perfectly matched to the forks' natural frequencies, and the forks are positioned correctly, their sound waves will create interference patterns that naturally stabilize the entire arrangement. The forks "want" to stay in that specific geometric relationship because that's where their powered oscillations create constructive rather than destructive interference.

This explains several mysterious aspects of molecular behavior:

Why molecules have specific shapes: Each arrangement of nuclei creates different substrate interference patterns. Only certain geometric arrangements create stable, constructive interference powered by the substrate's restorative pressure.

Why molecular bonds have specific strengths: The bond strength corresponds to how strongly the substrate resonance patterns reinforce each other in that particular geometric arrangement, all powered by the same restorative pressure that creates gravitational effects.

Why molecules vibrate at characteristic frequencies: The entire molecular structure is a substrate resonance system powered by equilibrium restoration around displaced nuclei. Like any resonant system, it has natural frequencies determined by the substrate's mechanical properties.

Why heat can break molecular bonds: Sufficient thermal energy can disrupt the delicate substrate resonance patterns by overwhelming the restorative pressure oscillations that maintain molecular geometry.

Why molecules are self-maintaining: Unlike human-made structures that require external energy for maintenance, molecular structures are continuously powered by the substrate's tendency to restore equilibrium around nuclear masses. The same mechanism that creates gravity automatically provides the energy that maintains molecular integrity.

The Hammer and Wood Analogy: Why Molecular Bonds Break While Nuclear Bonds Remain

This substrate hierarchy explains a fundamental observation that puzzles many students: why can we easily break apart a block of wood with a hammer, yet nuclear bonds remain completely unaffected?

The answer lies in understanding that we're operating in entirely different substrate displacement regimes. When you strike wood with a hammer, you're not challenging nuclear bonds at all - you're disrupting the delicate substrate resonance patterns that hold molecules together.

Think of it this way: molecular bonds are like a complex musical harmony being performed by an orchestra. The substrate oscillations around each nucleus create interference patterns that lock atoms into specific geometric arrangements, much like musicians following a conductor to create beautiful, coordinated music. A hammer blow is like suddenly shouting loudly in the concert hall - it provides enough disruptive energy to overwhelm the delicate frequency synchronizations that maintain the molecular "performance," causing the molecular structure to collapse.

The nuclear bonds, however, are like the instruments themselves. While the shouting can disrupt the music (break molecular bonds), it cannot destroy the violins, trumpets, and pianos (the nuclear structure within each atom). Those instruments - the nuclear pressure wells with their extreme $\varphi \approx 10^{-30}$ substrate displacement - require millions of times more energy to affect.

This is why mechanical forces easily break chemical bonds but leave atomic nuclei completely untouched. We're disrupting the gentle substrate resonance patterns between atoms while the intense substrate pressure wells within atoms remain unperturbed. The wood fibers separate because their molecular harmonies are destroyed, but every carbon, hydrogen, and oxygen nucleus emerges from the destruction completely unchanged.

The elegance of substrate mechanics reveals itself here: the same medium that creates nuclear binding through extreme displacement also enables molecular structure through gentle interference patterns, operating seamlessly across vastly different energy scales within a single, unified framework.

The Periodic Table as a Resonance Catalog

This perspective transforms the periodic table from a catalog of different elements into a catalog of different substrate resonance patterns. Each element represents a different way that substrate oscillations can create stable nuclear patterns.

Hydrogen, with its single proton, creates the simplest possible nuclear resonance pattern in the substrate. Helium, with two protons and two neutrons, creates a different but still simple pattern that happens to be extraordinarily stable (which is why helium is so chemically inert).

As we move through the periodic table to heavier elements, we encounter increasingly complex substrate resonance patterns. Some patterns are inherently stable (like carbon, which can form incredibly diverse molecular structures), while others are inherently unstable (like the radioactive elements that decay quickly).

The "magic numbers" in nuclear physics - specific numbers of protons and neutrons that create particularly stable nuclei - might correspond to substrate resonance patterns that create especially strong constructive interference. It's like finding the perfect chord progressions in music - certain combinations of oscillations naturally harmonize better than others.

Crystal Structure: Substrate Geometry Made Visible

Crystals provide some of the most beautiful examples of how substrate resonance patterns might create macroscopic structure. When atoms arrange themselves into perfect crystal lattices - creating gems, metals, and minerals with precise geometric patterns - they might be responding to substrate standing wave patterns that favor specific spatial arrangements.

Picture throwing sand onto a large drumhead and then driving the drumhead with multiple frequencies simultaneously. The sand would arrange itself into complex, repeating patterns determined by the interference between the different frequency components. Change the frequency mix, and you get completely different patterns.

Crystal growth might work similarly, with atoms settling into positions determined by the substrate resonance patterns created by the existing crystal structure. Each atom that joins the crystal finds the position where it adds constructively to the overall substrate interference pattern, creating the remarkably regular geometric arrangements we see in crystalline materials.

This explains:

Why crystals grow in specific geometric forms: The substrate resonance patterns naturally favor certain spatial arrangements.

Why crystal structure depends on temperature and pressure: These conditions affect the substrate resonance patterns.

Why some materials form multiple crystal structures: Different environmental conditions can stabilize different substrate resonance patterns.

Why crystals can have perfect internal order: The substrate standing wave patterns provide a template for atomic arrangement.

Biological Cymatics: The Resonance of Life

The most remarkable possibility is that biological structures - from DNA molecules to entire organisms - might be created and maintained by incredibly sophisticated substrate resonance patterns.

Consider the double helix structure of DNA. This elegant spiral arrangement holds the genetic information for all life, but what actually maintains its precise geometry? The substrate resonance model suggests that DNA might be a complex standing wave pattern in the substrate, with the nucleotide bases arranged in positions that create stable constructive interference.

This could explain several mysteries of biological structure:

Protein Folding: Proteins must fold into precise three-dimensional shapes to function properly, but how they achieve these shapes from linear amino acid sequences remains partly mysterious. If proteins are substrate resonance patterns, they would naturally fold into the geometric arrangements that create the most stable substrate interference patterns.

Cellular Structure: The complex internal organization of cells - organelles, membranes, molecular machinery - might represent stable substrate resonance patterns that naturally maintain themselves through substrate oscillations.

Developmental Biology: How does a single fertilized cell develop into a complex organism with precisely organized tissues and organs? If development follows substrate resonance templates, the growing organism would naturally organize itself according to substrate interference patterns encoded in its genetic material.

Healing and Regeneration: The body's remarkable ability to repair damage and regenerate tissue might involve restoring disrupted substrate resonance patterns to their stable configurations.

The Molecular Memory

One of the most intriguing implications of substrate resonance is that molecular structures might have a kind of "memory" encoded in substrate oscillation patterns. If a molecule's shape and properties are maintained by substrate standing waves, then these patterns might persist even when the original molecule is disrupted.

This could explain some puzzling phenomena:

Homeopathy: While controversial, some people report effects from extremely diluted solutions that shouldn't contain any active molecules. If the original molecules created persistent substrate resonance patterns, these patterns might remain in the solution even after the molecules are removed.

Water Memory: Some researchers have reported that water can somehow "remember" substances that were previously dissolved in it. If these substances created substrate resonance patterns in the water, the patterns might persist as standing waves in the substrate even after the substances are removed.

Phantom Limb Sensations: People who lose limbs sometimes continue to feel sensations in the missing body parts. If the body's structure is maintained by substrate resonance patterns, these patterns might persist even after the physical tissue is gone.

Molecular Engineering Through Resonance

Understanding molecular structure as substrate resonance patterns opens up revolutionary technological possibilities. Instead of manipulating molecules through crude chemical reactions, we might learn to create and modify molecular structures by controlling substrate oscillation patterns directly.

Designer Molecules: By creating specific substrate resonance patterns, we might be able to assemble molecules with any desired properties. Need a material that's lighter than aluminum but stronger than steel? Create the substrate resonance pattern that naturally arranges atoms into that configuration.

Programmable Materials: Materials could be designed with substrate resonance patterns that can be modified on command. Change the driving frequency, and the material changes its properties - now flexible, now rigid, now conductive, now insulating.

Molecular Repair: Damaged molecular structures could be repaired by restoring their original substrate resonance patterns. Instead of replacing broken molecules with new ones, we could simply regenerate the substrate oscillations that maintain the desired molecular geometry.

Crystal Engineering: We could grow crystals with perfect precision by creating substrate standing wave templates that guide atomic arrangement. Custom crystals for electronics, optics, and materials science could be designed and grown to exact specifications.

The Frequency Prescription

If biological structures are maintained by substrate resonance patterns, then medical treatment might eventually involve adjusting the substrate oscillation frequencies that maintain cellular and molecular organization.

Different diseases might correspond to disruptions in different substrate resonance patterns:

Cancer: Cells whose substrate resonance patterns have become destabilized, causing uncontrolled growth.

Genetic disorders: Disrupted substrate patterns that prevent normal protein folding or cellular organization.

Aging: Gradual decay in the substrate resonance patterns that maintain cellular structure and function.

Treatment could involve using precisely tuned substrate oscillations to restore healthy resonance patterns:

Resonance therapy: Using specific frequencies to reinforce disrupted substrate patterns.

Molecular tuning: Adjusting substrate oscillations to optimize cellular function.

Regenerative resonance: Using substrate patterns to guide tissue repair and regeneration.

The Observable Evidence

The cymatics connection isn't just theoretical - there are observable phenomena that might represent substrate resonance effects:

Molecular Vibration Spectra: Every molecule absorbs and emits light at characteristic frequencies that correspond to its vibrational modes. If molecules are substrate resonance patterns, these spectra might be direct observations of substrate oscillation frequencies.

Crystal Growth Patterns: The precise geometric arrangements of crystal formation might be visible manifestations of substrate standing wave patterns guiding atomic arrangement.

Biological Rhythms: The numerous rhythmic processes in living systems - heartbeats, brain waves, circadian cycles - might represent biological systems synchronizing with substrate oscillation patterns.

Morphic Resonance: Some researchers have reported that when animals learn new behaviors in one location, animals of the same species elsewhere seem to learn the same behaviors more easily.

The Experimental Path

Testing the substrate resonance model would require experiments designed to detect correlations between vibration and molecular stability:

Frequency Disruption Experiments: If molecules are maintained by substrate resonance, then precisely targeted electromagnetic fields at specific frequencies should be able to disrupt molecular bonds without thermal damage.

Resonance Enhancement Studies: Applying substrate oscillations at calculated frequencies should stabilize molecular structures and possibly enhance chemical reaction rates or selectivity.

Pattern Restoration Experiments: If substrate resonance patterns persist after molecular disruption, it should be possible to restore molecular structures by reapplying the correct oscillation frequencies.

Biological Frequency Studies: Testing whether specific electromagnetic frequencies can influence biological processes in ways that correspond to calculated substrate resonance patterns.

The Symphony of Structure

Perhaps the most beautiful aspect of understanding matter through substrate resonance is the musical metaphor it provides. Instead of random collections of particles held together by mysterious forces, we might have a vast, cosmic symphony playing out in the substrate medium.

Every atom vibrates at its characteristic frequencies.

Every molecule represents a harmony between multiple atomic oscillations.

Every crystal is a repeating rhythm in the substrate.

Every living organism is an incredibly complex musical composition maintaining itself through precisely coordinated resonance patterns.

You're not just made of carbon, hydrogen, oxygen, and nitrogen arranged in complex molecules. You might be a temporary symphony in the substrate medium - millions of oscillating patterns harmonizing together to create the resonance structure you experience as your body.

When you think, substrate patterns oscillate in your brain. When you move, substrate resonances adjust throughout your musculature. When you heal from injury, substrate patterns reorganize to restore the damaged symphony.

The physical world isn't held together by mysterious forces - it might be held together by the most elegant music ever composed, playing continuously in the invisible medium that fills all space.

And unlike the sand patterns on a vibrating plate, which disappear when the sound stops, the substrate resonance patterns that create matter might be self-sustaining - oscillations that maintain themselves through the fundamental properties of the substrate medium.

The universe might be singing itself into existence, one molecule at a time.

"If you want to find the secrets of the universe, think in terms of energy, frequency and vibration."

Tesla

Chapter 9

Inertia - The Resistance Revealed

Why Things Fight Back When You Try to Move Them

You're driving down the highway at a steady 70 miles per hour. The car feels perfectly smooth, as if you're sitting still. But then you floor the accelerator, and suddenly you're pressed back into your seat by an invisible force. Your coffee sloshes backward, loose objects slide toward the rear of the car, and you feel a distinct sensation of being pushed.

What just happened? Why does constant motion feel like no motion at all, while changing motion creates such dramatic effects?

Most physics textbooks will tell you it's "inertia" - the tendency of objects to resist changes in motion. But that's not really an explanation, it's just a name. It's like saying rocks fall because of "gravity" without explaining what gravity actually is.

The substrate model suggests that inertia isn't a mysterious property of matter - it's the substrate medium's resistance to being rearranged around accelerating objects. When you accelerate, you're forcing the substrate to reorganize its flow patterns around you, and the substrate pushes back. That pushback is what you feel as an inertial force.

The Fifth Fundamental Phenomenon

We've explored how substrate mechanics explains the four fundamental forces - gravity, electromagnetism, and the strong and weak nuclear forces. But there's a fifth phenomenon that's so universal we often forget it needs explaining: inertia.

Inertia affects every single piece of matter in the universe, from electrons to galaxies. It's what makes spacecraft need fuel, what throws you forward when a car brakes suddenly, and what keeps planets in their orbits. Yet for all its universality, inertia remains one of the least understood aspects of physics.

Newton described what inertia does but never explained why it works. Einstein connected inertia to gravity through the equivalence principle but didn't provide a mechanical explanation for either phenomenon.

The substrate model suggests that inertia and gravity are two aspects of the same underlying process: both emerge from how matter interacts with the substrate medium.

The Honey Analogy

To understand substrate inertia, imagine you're a small object moving through thick honey. When you move at constant velocity, the honey flows smoothly around you with minimal resistance. But when you try to accelerate - to change your speed or direction - something different happens.

Acceleration forces you to push the honey around you, creating high-pressure zones in front of your motion and low-pressure zones behind. The honey resists this forced rearrangement and pushes back, making acceleration much more difficult than constant motion. This is essentially what happens with substrate inertia. The substrate medium fills all space like invisible honey. Constant velocity motion allows smooth substrate flow with minimal resistance. But acceleration forces substrate rearrangement, and the substrate's resistance to this rearrangement is what we experience as inertial effects.

The Car Acceleration Experience

Let's analyze what happens during your highway acceleration in substrate terms:

Constant 70 mph: The substrate flows smoothly around you and the car in a stable pattern. You've reached equilibrium with the medium - no pressure gradients, no resistance, no forces felt. The substrate has "settled" into a steady flow configuration that moves with you.

Sudden acceleration: Now you're trying to change the substrate flow pattern rapidly. The substrate around you must rearrange itself to accommodate your new motion. This creates:

High pressure in the direction of acceleration (substrate gets compressed as you push into it).

Low pressure behind you (substrate gets stretched as you pull away from it).

Pressure gradient from high to low pressure across your body.

The substrate naturally tries to equalize these pressure differences by flowing from high pressure to low pressure. This substrate flow is what pushes you back into your seat - it's literally the substrate medium flowing around you to relieve the pressure imbalance created by your acceleration.

Why Constant Velocity Feels Like Nothing

This explains one of the most puzzling aspects of motion: why constant velocity - even very high velocity - feels exactly like being at rest.

Whether you're sitting still or cruising at 500 miles per hour in a smooth airplane, you feel no force because the substrate flow pattern around you is stable. There are no pressure gradients, no forced rearrangement, no substrate resistance.

Einstein recognized this as the principle of relativity: you can't detect uniform motion through any experiment performed inside your reference frame. In substrate terms, this makes perfect mechanical sense - uniform motion allows the substrate to flow in stable patterns that don't create any detectable effects.

Only acceleration or change of motion disturbs the substrate equilibrium and creates the pressure gradients we can feel as forces.

The Mathematics of Substrate Resistance

The substrate model provides a precise mathematical description of why objects resist acceleration. When you try to accelerate an object, you're forcing the substrate around it to change its flow pattern. This requires energy and creates resistance.

The resistance force depends on:

How fast you're trying to accelerate (higher acceleration = more forced rearrangement).

How much substrate needs rearranging (larger objects displace more substrate).

The substrate's resistance to changes (substrate viscosity).

This naturally gives us Newton's famous equation $F = ma$, but now we understand the mechanical reason behind it. The force needed to accelerate an object equals the mass times acceleration because mass represents how strongly an object couples to the substrate, and acceleration represents how much substrate rearrangement you're demanding.

The Scale of Substrate Effects

Just as with other substrate phenomena, inertial effects operate differently at different scales:

Everyday Objects (cars, people, aircraft): These create pressure gradients in the substrate during acceleration, but they don't create actual voids or cavitation zones. The substrate density might vary by 1-10% around an accelerating object, creating manageable pressure differences that we experience as normal inertial forces.

Extreme Acceleration (fighter jets, rockets, particle accelerators): Higher acceleration rates create steeper pressure gradients and stronger substrate flow effects. This is why high-g acceleration feels so intense - you're forcing more dramatic substrate rearrangement.

Relativistic Speeds: As objects approach the speed of light, substrate rearrangement becomes increasingly difficult. The substrate essentially becomes "stiffer" at high velocities, which is why it takes infinite energy to accelerate massive objects to light speed. Light speed represents the maximum rate at which the substrate can rearrange itself.

The Equivalence Principle Explained

Einstein's equivalence principle states that gravitational effects and inertial effects are indistinguishable. You can't tell the difference between being in a gravitational field and being accelerated in empty space.

The substrate model explains why this principle works: both gravity and inertia involve identical substrate pressure gradient effects.

Gravity: Large masses create static pressure gradients in the substrate that push objects toward areas of lower substrate density.

Inertia: Acceleration creates dynamic pressure gradients in the substrate that push back against the accelerating object.

Both involve substrate pressure pushing on objects - the only difference is whether the pressure gradients are static (gravity) or dynamic (inertia). From the object's perspective, substrate pressure is substrate pressure, regardless of what creates it.

Why We Don't Feel Earth's Motion

This also explains why we don't feel Earth's motion through space. Earth is spinning at 1,000 mph at the equator, orbiting the Sun at 67,000 mph, and racing through the galaxy at 514,000 mph. Yet we feel none of this motion.

The reason is that all these motions are essentially constant velocity (or very slowly changing) on human timescales. The substrate has settled into stable flow patterns around the Earth system. We're carried along in Earth's substrate flow pattern like leaves floating on a steadily moving river.

We only feel motion when there are rapid accelerations that disturb the local substrate equilibrium - like earthquakes, vehicle acceleration, or carnival rides.

Technological Implications

Understanding inertia as substrate resistance opens up remarkable technological possibilities:

Inertial Dampening: If we could manipulate substrate density around a spacecraft, we might be able to reduce inertial effects. Lower substrate density would mean less resistance to acceleration, allowing rapid acceleration with minimal force on passengers.

Advanced Propulsion: Instead of fighting against inertial forces, we might learn to work with substrate flow patterns. A propulsion system that creates appropriate substrate pressure gradients could accelerate smoothly without the harsh forces typical of rocket propulsion.

Gravitational Control: Since gravity and inertia involve the same substrate pressure mechanics, learning to control one might automatically give us control over the other. Artificial gravity systems might work by creating controlled substrate pressure gradients.

The Universal Experience

Perhaps the most beautiful aspect of understanding inertia through substrate mechanics is how it reveals the hidden universality of our experience. Every time you feel pressed back in an accelerating car, pushed forward in a braking vehicle, or thrown sideways around a sharp turn, you're directly experiencing the substrate medium's response to motion changes.

These aren't just mechanical effects - they're glimpses into the fundamental medium that makes all physics possible. The substrate pushback you feel during acceleration is the same medium that carries light waves, creates gravitational effects, and maintains the structure of matter itself.

You're not just a passenger in a vehicle accelerating through empty space. You're swimming in an invisible ocean of substrate that extends from your car to the farthest galaxies, and every acceleration you experience is that ocean pushing back as it rearranges itself around your motion.

Inertia isn't mysterious anymore. It's fluid dynamics in the most fundamental medium in the universe - and every time you feel it, you're experiencing proof that space isn't empty after all.

The Completion

With inertia explained through substrate mechanics, we've now accounted for all five fundamental phenomena that shape our physical experience:

Gravity: Static pressure gradients in the substrate.

Electromagnetism: Circulation patterns in the substrate.

Strong nuclear force: Extreme substrate compression effects.

Weak nuclear force: Substrate instability and oscillations.

Inertia: Dynamic substrate resistance to flow rearrangement.

Every force you've ever experienced, every resistance you've ever felt, every motion you've ever detected - all emerge from the mechanical properties of the substrate medium that fills space. Physics becomes comprehensible again when we recognize the invisible ocean we're swimming in.

"Every body perseveres in its state of rest, or of uniform motion in a right line, unless it is compelled to change that state by forces impressed thereon."

Newton — Isaac Newton (Principia, First Law)

CHAPTER 10

THE GREEN PAINT PROBLEM

WHEN PERFECT MATH HIDES THE MECHANISM

Imagine you're a painter who has always used green paint. For your entire career, you've bought tubes labeled "green paint" from the art store, and this green has worked perfectly for everything you want to paint. Your green landscapes are gorgeous, your technique is flawless, and everyone admires your artistic skill.

You've used green paint for so long that you think of it as a fundamental color. You write papers about "green paint theory," develop complex mathematical equations describing how green paint behaves under different lighting conditions, and build an entire artistic philosophy around the intrinsic properties of greenness.

Then one day, someone suggests that your "green paint" might actually be a mixture of yellow and blue that someone else combined before putting it in the tube. Your immediate reaction is dismissal: "Impossible! I've used green paint for decades. It works perfectly. Every painting I've ever created proves that green is a fundamental color!"

But here's the thing: you're absolutely right that green paint works perfectly. Every success you've achieved is real. The mathematical equations describing green paint behavior are completely accurate. The only thing you missed is that green paint's success comes from underlying yellow and blue components that you never suspected were there.

This is the Green Paint Problem in physics - and it might explain the relationship between Einstein's incredibly successful equations and the substrate mechanics we've been exploring.

The 200-Year Thermodynamics Gap

This exact scenario has happened before in physics, and the historical parallel is illuminating. From the 1660s to the 1860s - for over 200 years - scientists had perfectly working mathematical laws for heat, pressure, and temperature without having any idea what heat, pressure, and temperature actually were.

The Mathematical Laws Came First (1660s-1850s):

Robert Boyle discovered in 1662 that pressure times volume equals a constant (PV = constant). Jacques Charles found in 1787 that volume divided by temperature stays constant (V/T = constant). Scientists combined these into the ideal gas law, added the laws of thermodynamics, and could predict the behavior of steam engines, refrigerators, and heat transfer with extraordinary precision.

Engineers built the Industrial Revolution using these laws. Steam trains crossed continents, factories powered entire cities, and thermodynamic calculations proved accurate to many decimal places. The mathematics was not just successful - it was transformational for human civilization.

But Nobody Knew Why It Worked:

For two centuries, scientists had no idea what temperature actually was. They knew it followed mathematical relationships with pressure and volume, but the physical mechanism remained completely mysterious. Heat was thought to be a fluid called "caloric" that flowed between objects. Pressure was just a mathematical parameter in equations. Temperature was simply whatever thermometers measured.

The mathematical formalism was perfect "green paint" - it worked beautifully for every practical purpose, but nobody realized it was actually a mixture of simpler, more fundamental components.

The Mechanism Finally Revealed (1860s-1900s):

Then James Clerk Maxwell, Ludwig Boltzmann, and others developed kinetic theory and revealed the "yellow and blue" components underneath thermodynamics:

Temperature isn't a fundamental quantity - it's the average kinetic energy of countless moving molecules.

Pressure isn't a basic force - it's the cumulative effect of billions of molecular collisions with container walls.

Heat isn't a fluid - it's the transfer of kinetic energy between molecular populations.

Entropy isn't mysterious - it's a measure of how many ways molecules can be arranged.

Suddenly, the mathematical laws that had worked perfectly for 200 years had mechanical explanations. The "green paint" of thermodynamics turned out to be the "yellow and blue" of molecular motion.

Einstein's Green Paint

This historical precedent raises a fascinating possibility: What if Einstein's equations are the "green paint" of fundamental physics?

Einstein's field equations work with extraordinary precision. They predict the motion of planets, the behavior of GPS satellites, the bending of light around massive objects, and the existence of gravitational waves. Just like thermodynamic equations, Einstein's mathematics is not just successful - it's transformational for our understanding and technology.

But Einstein's equations describe *what* happens (spacetime curvature) without necessarily explaining *why* it happens. They tell us that massive objects curve spacetime and that this curvature manifests as gravitational effects. But they don't explain the physical mechanism by which matter creates curvature or how curvature generates forces.

What if Einstein discovered mathematical relationships that work perfectly because they describe the large-scale effects of substrate mechanics - just like thermodynamic laws perfectly described the large-scale effects of molecular motion?

The Substrate "Yellow and Blue":

In this view:

Spacetime curvature isn't fundamental - it's the mathematical description of substrate pressure gradients.

Gravitational fields aren't basic entities - they're the cumulative effect of substrate displacement around massive objects.

The speed of light limit isn't about spacetime geometry - it's the maximum rate of substrate rearrangement.

Time dilation effects aren't about the nature of time - they're about how substrate dynamics affect the rate of physical processes.

Just as kinetic theory revealed that thermodynamic "forces" were actually statistical effects of molecular motion, substrate theory might reveal that gravitational "forces" are actually mechanical effects of substrate displacement.

Why This Doesn't Diminish Einstein

Understanding green paint as yellow plus blue doesn't make your paintings any less beautiful or your artistic technique any less skillful. Similarly, understanding

Einstein's equations as descriptions of substrate effects doesn't diminish their mathematical elegance or practical success.

Einstein discovered genuine mathematical relationships that accurately describe observable phenomena. His equations work because they correctly capture the large-scale behavior of substrate mechanics, just as thermodynamic equations work because they correctly capture the large-scale behavior of molecular motion.

The substrate model doesn't replace Einstein's equations - it potentially explains why they work so well. Instead of mysterious spacetime curvature, we get mechanical substrate displacement. Instead of incomprehensible gravitational fields, we get understandable pressure gradients. Instead of abstract mathematical relationships, we get visualizable physical processes.

The Pattern of Scientific Progress

This pattern repeats throughout the history of physics:

Kepler discovered mathematical laws describing planetary motion (elliptical orbits, equal areas in equal times) without knowing why planets followed these patterns.

Newton later revealed the mechanical forces that created Kepler's mathematical relationships.

Maxwell discovered mathematical equations describing electromagnetic phenomena without knowing what electric and magnetic fields actually were.

Later physicists revealed that electromagnetic fields could be understood as excitations in quantum field theory.

In each case, mathematical success came first, often by decades or centuries, followed eventually by mechanical understanding of the underlying physical processes.

The GPS Satellite Test

Einstein's equations are so accurate that GPS satellites must include relativistic corrections to maintain precision. Without these corrections, GPS would drift by miles per day. This is often cited as definitive proof that Einstein's spacetime description is fundamentally correct.

But consider the thermodynamics analogy: Steam engines worked perfectly using thermodynamic equations for 200 years before anyone understood molecular motion. The practical success of thermodynamics didn't prove that "caloric fluid" was real - it proved that the mathematical relationships correctly described the large-scale effects of underlying molecular processes.

Similarly, GPS accuracy might prove that Einstein's equations correctly describe the large-scale effects of substrate mechanics without proving that spacetime curvature is fundamental. The substrate displacement created by Earth's mass might affect the rate of atomic processes in GPS satellite clocks, creating the time dilation effects that Einstein's equations predict.

The mathematical corrections work perfectly because they accurately account for substrate effects, just as thermodynamic calculations worked perfectly because they accurately accounted for molecular statistical effects.

The Experimental Challenge

If Einstein's equations are "green paint" describing substrate effects, how would we tell the difference? This is actually more subtle than it might seem.

Thermodynamic equations remained unquestioned for centuries because molecular motion was too small to observe directly. Only when technology advanced enough to study individual molecular behavior could scientists confirm the kinetic theory.

Similarly, substrate effects might be too subtle to detect with current technology. The substrate displacement that creates gravitational effects operates on such vast scales and with such gentle gradients that we might only observe its large-scale consequences - which Einstein's equations describe perfectly.

But just as molecular theory made predictions that eventually distinguished it from pure thermodynamics, substrate theory makes predictions that might eventually distinguish it from pure relativity:

Substrate oscillation signatures in high-energy particle collisions.

Substrate wave propagation faster than electromagnetic radiation.

Substrate pressure variations in strong gravitational fields.

Substrate circulation effects in electromagnetic phenomena.

The Beauty of Deeper Understanding

The most exciting aspect of the substrate model isn't that it challenges Einstein's success, but that it might explain Einstein's success through mechanical processes we can visualize and understand.

Kinetic theory didn't make thermodynamics less useful - it made thermodynamics more beautiful by revealing the elegant dance of molecules that creates thermal behavior. Similarly, substrate mechanics might make Einstein's equations more beautiful by revealing the elegant flow of substrate that creates gravitational effects.

Instead of spacetime mysteriously curving in response to matter, we might have the substrate mechanically displaced by matter, creating pressure gradients that guide the motion of other matter. Instead of incomprehensible gravitational fields, we might have comprehensible fluid dynamics in the medium that fills space.

Physics doesn't lose its mathematical precision - it gains mechanical clarity.

The Question of Timing

How long might it take to definitively confirm substrate mechanics? The thermodynamics precedent suggests that mathematical formalism can remain unquestioned for centuries before the underlying mechanism is revealed.

But modern experimental techniques might accelerate this timeline. We have particle accelerators that can probe substrate effects directly, gravitational wave detectors that might reveal substrate oscillations, and precision instruments that could detect substrate pressure variations.

The challenge isn't necessarily technological - it's conceptual. Scientists need to start looking for substrate effects rather than assuming that Einstein's mathematical success proves spacetime geometry is fundamental.

The Green Paint Lesson

The thermodynamics precedent teaches us that mathematical success, however spectacular, doesn't necessarily reveal fundamental physical mechanisms. Perfect "green paint" can work for centuries before anyone discovers the "yellow and blue" components underneath.

Einstein's equations might be the most successful "green paint" in the history of physics - mathematical relationships that work with extraordinary precision because they accurately describe the large-scale effects of substrate mechanics that nobody has recognized yet.

This doesn't diminish Einstein's achievement any more than molecular theory diminishes the achievement of thermodynamics pioneers. It simply suggests that there might be an even deeper level of understanding waiting to be discovered - a level where forces become flows, where fields become fluid dynamics, and where the mysterious becomes mechanical.

This reveals something profound about the nature of scientific "understanding":

Mathematical success ≠ Physical understanding

Practical applications ≠ Fundamental mechanism

Predictive power ≠ Actual comprehension

The substrate universe might be the "yellow and blue" that explains why Einstein's "green paint" works so beautifully. And if so, we're on the verge of the most exciting discovery in physics since kinetic theory revealed the molecular reality behind thermodynamics.

The green paint is perfect. But imagine how much more beautiful it becomes when you finally see the colors underneath.

"I have a friend who's an artist and has sometimes taken a view which I don't agree with. He'll hold up a flower and say 'look how beautiful it is,' and I'll agree. Then he says 'I as an artist can see how beautiful this is but you as a scientist take this all apart and it becomes a dull thing,' and I think that he's kind of nutty... I can appreciate the beauty of a flower. At the same time, I see much more about the flower than he sees."

Feynman

CHAPTER 11

FASTER THAN LIGHT (BUT NOT BREAKING EINSTEIN?)

THE SECRET HIGHWAY ABOVE THE SPEED LIMIT

Einstein's speed limit is absolute: nothing can travel faster than light. This cosmic speed limit of 186,282 miles per second isn't just a suggestion - it's built into the fundamental structure of spacetime itself. Matter, energy, and information are all supposedly bound by this universal constraint.

But what if Einstein's speed limit only applies to one layer of reality? What if there's a deeper level - the substrate medium itself - that operates by different rules entirely?

The substrate model suggests something remarkable: while matter and energy are indeed limited to light speed, the substrate medium that carries forces and fields might support waves that travel billions of times faster than light. These "substrate waves" could provide instantaneous communication across any distance without violating relativity, because they don't carry matter or energy - they carry pure information through the medium that makes physics possible.

The Two-Highway Universe

To understand how faster-than-light substrate waves can coexist with Einstein's speed limit, imagine the universe as having two separate highway systems:

Highway One: The Matter-Energy Interstate This is the highway system that Einstein described. Matter, energy, and everything made of particles must travel on this highway system, and the absolute speed limit is the speed of light.

No exceptions. This speed limit exists because matter and energy interact with the substrate in ways that create resistance - like trying to push objects through thick honey.

Highway Two: The Substrate Superhighway

This is a faster highway system that exists within the substrate medium itself. Information can travel on this highway system at speeds that dwarf the speed of light - but only pure information, not matter or energy. It's like having pneumatic tubes that can instantly send messages through a building, even though people still have to walk through the hallways at normal speeds.

The key insight is that these two highway systems don't interfere with each other. Matter and energy are stuck on the slower highway, bound by Einstein's speed limit. But information can use the faster highway, traveling through the substrate medium itself at speeds limited only by the mechanical properties of the substrate.

The Speed of Substrate Waves

How fast might substrate waves travel? The speed of any wave in a medium depends on two properties: the medium's stiffness (how strongly it resists being deformed) and its density (how much inertia it has).

For sound waves in air, this gives a speed of about 767 miles per hour. For electromagnetic waves in the substrate (light), this gives a speed of about 670 million miles per hour. But for pressure waves propagating through the substrate medium itself - substrate waves - the calculation yields something extraordinary.

The substrate appears to be incredibly stiff (it supports forces across cosmic distances) but incredibly light (we can barely detect its presence). When you divide a very large stiffness by a very small density, you get a very large speed.

Current calculations suggest substrate waves might travel at approximately 3.6 × 10^34 meters per second - about 100 billion billion times faster than light. At this speed, a substrate wave could travel from Earth to the nearest star in about 0.0000004 seconds, or from one end of our galaxy to the other in about 40 minutes.

These aren't electromagnetic waves (which are limited to light speed) or matter waves (which are limited to even slower speeds). They're mechanical waves in the substrate medium itself - compression and expansion waves that propagate through the "fabric" of space at speeds that make light seem to crawl.

Quantum Entanglement: The Substrate Connection

This faster-than-light substrate wave propagation might finally explain one of the most mysterious phenomena in physics: quantum entanglement. When two particles become "entangled," measuring one particle instantly affects the other, no matter how far apart they are. Einstein called this "spooky action at a distance" and refused to believe it was real.

But experiments have proven that entanglement is absolutely real. Measure the spin of one particle, and its entangled partner immediately "knows" what result you got, even if it's on the other side of the galaxy. The effect is truly instantaneous - faster than light could travel between the particles.

The substrate model provides a beautifully mechanical explanation: entangled particles aren't mysteriously connected across empty space. They're connected by standing wave patterns in the substrate medium - persistent interference patterns that maintain correlation between distant locations.

Imagine two corks floating on opposite sides of a lake, connected by a standing wave pattern on the water's surface. If you push down on one cork, the wave pattern instantly adjusts throughout the entire lake, and the other cork immediately responds. The "information" that you pushed the first cork travels across the lake at the speed of water waves - much faster than anything floating on the surface could travel.

Similarly, entangled particles might be connected by standing wave patterns in the substrate. When you measure one particle (essentially "pushing" on one part of the pattern), the substrate wave pattern instantly adjusts throughout space, and the other particle immediately responds. The correlation information travels through the substrate at substrate wave speeds - much faster than light, but without violating relativity because no matter or energy is being transported.

Why This Doesn't Break Causality

The crucial point is that substrate waves don't transport matter or energy - they only transport correlation information. You can't use them to send a message that violates causality (like warning yourself about something before it happens) because they don't carry energy that could be used to influence matter directly.

Think of it like this: if you have two synchronized clocks, you can instantly "know" what time it is on the other clock by looking at your local clock. The correlation information is instantaneous, but you can't use this to send messages or influence events.

The substrate wave correlations work similarly - they maintain instantaneous correlations between distant locations, but they can't be used to transmit energy or create causal paradoxes.

Einstein's relativity remains perfectly intact for everything made of matter and energy. But pure information - correlation patterns in the substrate medium - might operate by different rules entirely.

The Quantum Computing Revolution

If substrate waves really exist and can be controlled, they could revolutionize quantum computing. Current quantum computers are incredibly fragile because quantum states are easily disrupted by environmental interference. But if quantum correlations are maintained by substrate wave patterns, then quantum computers based on substrate engineering might be far more stable and powerful.

Imagine quantum computers that could maintain entanglement across any distance, process information at substrate wave speeds, and perform calculations using the interference patterns of substrate waves themselves. Such computers might be capable of simulations and calculations that dwarf anything possible with conventional electronics.

Instantaneous Communication

The most exciting possibility is that substrate waves might enable true instantaneous communication across any distance. If we could learn to generate and detect controlled substrate wave patterns, we could send information faster than light without violating any physical laws.

Picture a communication system based on substrate wave interference patterns. A transmitter on Earth could create specific patterns in the local substrate, and these patterns would propagate at substrate wave speeds to a receiver anywhere in the universe. The receiver could detect the patterns and decode the information, achieving communication that's essentially instantaneous across cosmic distances.

Such a communication system would transform space exploration. Instead of waiting years for radio signals to travel between Earth and distant spacecraft, mission controllers could communicate with interstellar probes in real time. Colonies on distant planets could maintain instantaneous contact with Earth. The solar system would become as connected as a modern city.

Detecting Substrate Waves

The challenge, of course, is that substrate waves might be incredibly difficult to detect directly. Just as electromagnetic waves were invisible to humans until we invented radio receivers, substrate waves might require entirely new types of detection equipment.

But there are promising approaches. If substrate waves create tiny changes in the local properties of space - affecting things like the speed of light or the strength of gravitational fields - then sensitive instruments might be able to detect these changes. Atomic clocks, laser interferometers, and quantum sensors might all be capable of detecting substrate wave signatures if we know what to look for.

Some researchers are already searching for such effects in existing experimental data. Tiny, unexplained variations in atomic clock rates, unexpected correlations in quantum experiments, and anomalous signals in gravitational wave detectors might all be evidence of substrate wave activity that we haven't recognized yet.

The Tesla Connection

Remarkably, some of the earliest hints about faster-than-light substrate waves might have come from Nikola Tesla's experiments over a century ago. Tesla claimed to have achieved wireless energy transmission and reported detecting signals that seemed to travel faster than light through the Earth itself.

Tesla's "longitudinal waves" and his theories about wireless power transmission bear striking resemblances to substrate wave predictions. Tesla might have been the first person to accidentally generate and detect substrate waves, though he lacked the theoretical framework to understand what he had discovered.

If Tesla's reports were accurate, then substrate wave technology might be much more accessible than we imagine. Instead of requiring exotic equipment and enormous energies, substrate wave generation might be achievable with relatively simple electrical devices - once we understand the principles involved.

I could elaborate on the Tesla Connection in great detail, however we will save this body of work as a dedicated and detailed publication.

The Navigation Revolution

Faster-than-light substrate waves could also revolutionize navigation and positioning. Current GPS systems rely on carefully timed radio signals from satellites, but these signals travel at light speed and can be blocked or interfered with easily.

A substrate wave positioning system could provide instantaneous, unblockable position information anywhere in the solar system. Spacecraft could navigate with perfect accuracy even in the depths of space, and the system would work regardless of electromagnetic interference or physical obstacles.

The Scientific Revolution

Perhaps most importantly, controllable substrate waves would transform scientific research itself. We could create "substrate telescopes" that use substrate wave interactions to study distant objects in ways impossible with electromagnetic radiation. We could probe the interior structure of planets, stars, and black holes by analyzing how they affect substrate wave propagation.

Substrate wave astronomy might reveal entirely new classes of cosmic phenomena - objects and events that don't emit electromagnetic radiation but do create detectable patterns in the substrate medium. The universe might be full of substrate wave sources that we've never even imagined.

The Speed of Thought

There's even a tantalizing possibility that biological systems might have evolved to use substrate waves. If consciousness involves quantum processes in the brain (as some theories suggest), then thoughts themselves might propagate through substrate wave patterns.

This could explain some of the mysterious aspects of consciousness - how the brain achieves unified awareness despite consisting of billions of separate neurons, how some people seem to have intuitive insights that transcend normal information processing, and why meditation and other practices that quiet electromagnetic brain activity sometimes seem to enhance awareness.

If thought operates partially through substrate waves, then understanding and controlling these waves might lead to technologies that interface directly with consciousness itself.

The Elegant Solution

The beauty of faster-than-light substrate waves is how they resolve the apparent conflict between quantum mechanics and relativity. Einstein's speed limit remains absolute for matter and energy under normal substrate conditions. Quantum entanglement becomes a perfectly mechanical process involving wave propagation in a medium. Information can travel faster than light without creating causal paradoxes.

Instead of having to choose between Einstein and quantum mechanics, we get both - each operating in their appropriate domain. Relativity governs the world of matter and energy. Substrate mechanics governs the world of correlations and information patterns.

The universe becomes more elegant, not more mysterious. Faster-than-light effects become as understandable as sound waves in air, and the apparent impossibility of instantaneous action at a distance dissolves into the simple mechanics of wave propagation in a medium.

We're not breaking Einstein's speed limit. We're discovering that reality has express lanes that Einstein never knew existed.

"Time and space are modes by which we think and not conditions in which we live."

Einstein

CHAPTER 12

THE QUANTUM CONNECTION

WHEN MYSTERY BECOMES MECHANICS

Walk into any physics classroom where quantum mechanics is being taught, and you'll witness one of the most peculiar rituals in modern education. Students are told to abandon everything they know about how the world works. They must accept that particles can be in multiple places at once, that measuring something changes its fundamental nature, and that the quantum world operates by rules completely different from everyday experience.

"Don't try to visualize it," professors warn. "Don't ask what's 'really' happening. Just learn the mathematics and accept that nature is fundamentally mysterious at the smallest scales."

But what if this entire approach is backwards? What if quantum mechanics seems mysterious only because we've been thinking about it with a crucial piece missing? What if every "weird" quantum effect becomes perfectly normal and visualizable once you recognize that space is filled with a substrate medium?

The substrate model suggests that quantum mechanics isn't mysterious at all. It's simply field mechanics applied to the most subtle medium in the universe - as understandable as ripples on a pond or pressure waves in air.

The Great Quantum Deception

For nearly a century, physics students have been told that quantum mechanics reveals the fundamental weirdness of reality. Particles can supposedly exist in multiple places simultaneously. The act of observation creates reality rather than simply revealing it. Things can influence each other instantly across vast distances without any connection between them.

This has created a peculiar situation where the deeper you study physics, the less sense the world seems to make. Students start with intuitive mechanics - objects have definite positions, causes precede effects, things interact through contact or identifiable connections. But as they advance, they're systematically trained to abandon this mechanical understanding and accept mathematical abstractions that defy visualization.

But what if the problem isn't that nature is weird - what if the problem is that we've been missing the most important component of the picture?

The False Duality Exposed

The central "mystery" of quantum mechanics is wave-particle duality - the idea that electrons, photons, and other quantum entities somehow behave as both waves and particles depending on how you observe them. This seems impossible because our everyday experience tells us that things are either waves (like ripples on water) or particles (like marbles) - they can't be both.

Here's the remarkable truth: our everyday intuition is absolutely correct. Wave-particle duality doesn't exist. It's a false concept that arose because physicists didn't recognize the substrate medium that fills space.

The substrate model reveals what's actually happening:

Particles are always particles - they're stable, localized field patterns in the substrate medium, like persistent vortices in a flowing river that maintain their structure as they move.

Waves are always waves - they're energy disturbances propagating through the substrate medium, like sound waves moving through air or ripples spreading across water.

There is no duality - every phenomenon is either a particle (a stable substrate field pattern) or a wave (energy propagating through substrate), never both simultaneously.

What we call an "electron" is always a particle - a stable substrate field configuration that maintains its pattern as it moves through the medium. What we call "light" is always a wave - electromagnetic energy propagating through the substrate at the medium's characteristic wave speed.

The confusion arose because physicists assumed particles were moving through empty space. Without recognizing the substrate medium, they couldn't explain how waves could exist or how particles could exhibit wave-like behavior. So they invented the impossible concept of wave-particle duality to paper over their lack of understanding about the medium.

The Double-Slit Experiment Demystified

The famous double-slit experiment has been used for decades to "prove" wave-particle duality, but it actually proves the opposite once you understand substrate mechanics.

Here's what actually happens: When you shine light through two parallel slits, you get an interference pattern on the screen - alternating bright and dark bands. This is exactly what you'd expect from waves propagating through a medium. Light consists of electromagnetic waves in the substrate, and when parts of the wave go through both slits simultaneously, they interfere with each other on the far side, just like water waves would.

The confusion arose when physicists tried the experiment with what they thought were "particles" - individual electrons, for example. Even when sending electrons through one at a time, they still got interference patterns. This seemed impossible: how could a single particle interfere with itself?

The substrate model makes this perfectly clear: electrons aren't particles in the sense of tiny solid balls bouncing around in empty space. They're stable field patterns in the substrate medium. An electron's field pattern can extend spatially through the substrate, just like a whirlpool in water can be localized but still involve water motion over a extended area.

When an electron field pattern encounters the double slits, parts of its substrate field extend through both slits simultaneously - just like a water wave would. These field components then interfere with each other, creating the interference pattern. But when the electron field pattern interacts with a detector screen, it creates a localized response at a specific point, which appears "particle-like."

When you try to detect "which slit the electron went through," you have to interact with the electron's substrate field pattern using detection equipment. This interaction necessarily disturbs the delicate field pattern, destroying the interference just like sticking a probe into a water wave disrupts the wave's coherent pattern.

The electron doesn't "know" you're watching it. The substrate field pattern simply can't maintain its extended coherent structure when you disrupt it with measurement interactions. There's no mystery - it's basic field mechanics in a physical medium.

Quantum Superposition: Extended Field Patterns

One of the most puzzling concepts in quantum mechanics is superposition - the idea that particles can exist in multiple states simultaneously until they're measured. Schrödinger's famous cat is supposedly both dead and alive until someone opens the box to check.

But in the substrate model, there's no mystery at all. What physicists call "superposition" is simply an extended substrate field pattern that covers multiple possible detection locations.

Think of throwing a large stone into the center of a calm pond. The resulting waves spread out in all directions, creating a complex pattern across the entire pond surface. At any moment, the water is disturbed at many different locations simultaneously. You could place detectors at various points around the pond, and each would register wave activity - but the wave isn't "in multiple states," it's simply an extended pattern in the water medium.

Similarly, when a quantum system is in "superposition," the particle's substrate field pattern extends over multiple locations. The field has definite, real structure throughout space - it's not "in multiple states," it's in one extended field configuration that happens to cover multiple possible measurement locations.

When you make a measurement, you're essentially asking "what's the substrate field doing at this specific location?" The answer you get depends on the field amplitude at that detection point. After measurement, you're no longer dealing with the full extended field pattern, but only with the localized part that your detector accessed.

This is why "superposition" appears to "collapse" upon measurement - the extended field pattern becomes localized through its interaction with the measurement apparatus.

The Uncertainty Principle: Basic Field Physics

The Heisenberg uncertainty principle states that you can't simultaneously know both the exact position and exact momentum of a particle. This has been interpreted as a fundamental limit on knowledge itself, as if the universe somehow prevents us from knowing certain combinations of information.

But the substrate model reveals it as nothing more than basic field physics - the same limitations that apply to any wave phenomenon in any medium.

Try to measure the exact location of a wave pulse on a rope. To do this, you need a very short, sharp pulse - essentially a spike. But sharp pulses necessarily contain a wide range of frequencies mixed together (this is basic wave mathematics). Since the speed of the pulse depends on its frequency content, a sharp pulse with many frequency components doesn't have a well-defined speed.

Conversely, if you want a wave pulse with a very precise speed, you need a pulse with a very narrow range of frequencies - essentially a pure tone. But pure tones extend over long distances, so you can't pinpoint exactly where the pulse is located.

This is exactly the uncertainty principle. Position corresponds to how localized the substrate field pattern is. Momentum corresponds to how well-defined the field's wave characteristics are. You can't have both simultaneously because that's simply not how field patterns work in any medium - it's basic wave physics, not a mysterious quantum effect.

Quantum Tunneling: Field Leakage

Quantum tunneling allows particles to pass through barriers that should be impenetrable according to classical physics. An electron can somehow "tunnel" through an electrical barrier even when it doesn't have enough energy to go over the barrier.

This seems to violate energy conservation and common sense. How can something pass through a solid wall without breaking the wall or gaining extra energy?

The substrate model makes tunneling perfectly intuitive. When a substrate field pattern encounters a barrier, most of the field reflects back, but some components can "leak" through the barrier if it's not infinitely thick or impenetrable.

This is exactly how field leakage works in any medium. Shout at a thick wall, and most of the sound reflects back, but some sound waves leak through and can be heard (faintly) on the other side. The thicker the wall, the less sound gets through, but there's always some leakage unless the barrier is infinitely thick.

Similarly, when an electron's substrate field pattern encounters a potential energy barrier, most of the field reflects back, but some field amplitude can leak through to the other side of the barrier. The probability of detecting the electron on the far side depends on how much field amplitude makes it through - just like how much sound makes it through a wall depends on the wall's thickness and the sound's intensity.

The electron doesn't mysteriously teleport through the barrier. Its substrate field simply has some amplitude on the far side of the barrier, giving a probability that measurements will detect the electron there.

Quantum Entanglement: Correlated Field Patterns

We've already seen how faster-than-light substrate waves can explain quantum entanglement through correlated field patterns in the substrate medium. But there's another important aspect to consider: why do particles become entangled in the first place?

The answer lies in how substrate field patterns interact when particles are created together or brought into close proximity. When two electrons are created simultaneously (like photon pairs from an excited atom) or when they interact closely, their substrate field patterns become synchronized - like two tuning forks that start vibrating in harmony.

These correlated field patterns maintain their relationship even when the particles separate, because the correlation is maintained by substrate wave connections that extend throughout space. It's not mysterious action at a distance - it's synchronized field oscillations in a physical medium, maintained by the medium's own wave propagation.

When you measure one particle, you're detecting the state of one part of the correlated field pattern. The substrate waves that maintain the correlation instantly adjust the pattern throughout space, immediately affecting the other particle's field configuration. The "information" travels through substrate waves at substrate wave speeds - much faster than light, but through a perfectly mechanical process.

The Measurement Problem Solved

The "measurement problem" in quantum mechanics asks: why does measurement seem to change quantum systems so dramatically? Before measurement, a system exists in "superposition" (extended field patterns). After measurement, it's in a definite state (localized field pattern). What causes this change?

The substrate model provides a clear, mechanical answer: measurement requires physical interaction between the quantum system's substrate field and the measuring device. This interaction necessarily involves energy transfer and field disturbance, which disrupts the delicate extended field patterns that create quantum effects.

It's exactly like trying to measure the complex interference pattern created by multiple overlapping water waves. To make the measurement, you have to stick instruments into the water, which inevitably disturbs the very wave patterns you're trying to measure. The more precise your measurement, the more you disrupt the field patterns.

The "collapse" isn't mysterious - it's the inevitable result of disturbing extended field patterns with localized measurement interactions. Any measurement process energetic enough to provide information is energetic enough to alter the field patterns being measured.

Quantum Computing: Field Pattern Processing

Understanding quantum mechanics as substrate field mechanics transforms how we think about quantum computing. Instead of mysterious "quantum bits" that exist in multiple states simultaneously, we have substrate field patterns that can store and process information through field interference effects.

Quantum computers work by creating controlled field patterns in the substrate and then manipulating these patterns through carefully designed interactions. The computational power comes from the ability to process many different field interference patterns simultaneously - like analyzing multiple overlapping wave patterns in a pond at the same time.

This perspective suggests new approaches to quantum computing based on direct substrate field manipulation rather than trying to isolate individual quantum particles. Such computers might be more stable and powerful because they would work with the substrate medium directly rather than fighting against environmental field interactions.

The End of Quantum Mystery

Perhaps the most satisfying aspect of understanding quantum mechanics through substrate field mechanics is how it restores intuitive comprehension to physics. Every "weird" quantum effect becomes a normal field phenomenon in a physical medium:

Wave-particle duality → False concept; particles are field patterns, waves are field propagation

Superposition → Extended field patterns covering multiple locations

Uncertainty principle → Basic limitations of field localization in any medium

Quantum tunneling → Field pattern leakage through barriers

Entanglement → Correlated field patterns maintained by substrate waves

Measurement collapse → Disruption of extended field patterns by measurement interactions

Instead of having to memorize mysterious rules that violate common sense, you can understand quantum behavior by thinking about field patterns in a physical medium. Instead of accepting that the quantum world is fundamentally incomprehensible, you can visualize quantum processes as clearly as watching ripples interact on a pond.

The revelation is profound: quantum mechanics isn't weird because nature is weird. Quantum mechanics seemed weird because we didn't recognize the substrate medium that makes it all work mechanically.

The Return of Physical Reality

The substrate model returns physics to physical reality. Instead of abstract mathematical entities existing in some non-physical realm, we have concrete field patterns in a real medium. Instead of measurements "creating reality," we have measurements detecting pre-existing field configurations. Instead of mysterious non-local connections, we have mechanical wave propagation through a physical substrate.

This doesn't diminish the precision or power of quantum mathematics - the equations remain just as accurate for calculating quantum phenomena. But now we understand what mathematics actually describes: field dynamics in the substrate medium rather than abstract probability relationships in mathematical space.

Quantum mechanics becomes a branch of substrate field theory, as mechanically understandable as any other area of physics. The quantum world isn't separate from the classical world - it's the same substrate medium operating at different scales and field intensities.

Quantum Understanding

Once you see the substrate, quantum mechanics becomes as intuitive as feeling pressure waves from a loudspeaker or watching interference patterns where water waves meet. It's all field dynamics in a physical medium - and there's nothing mysterious about that.

The electrons in your body, the photons illuminating this page, the quantum processes powering your electronic devices - they're all field patterns and wave propagation in the substrate medium that fills space. Every quantum phenomenon you've ever heard about is actually a perfectly normal field effect in the most subtle medium ever discovered.

Quantum mechanics isn't weird anymore. Its substrate field mechanics applied to very small scales and very subtle field patterns. And once you understand the medium, it becomes more beautiful and elegant than anyone ever imagined.

"We live on an island surrounded by a sea of ignorance. As our island of knowledge grows, so does the shore of our ignorance."

Wheeler

CHAPTER 13

BLACK HOLES - THE ULTIMATE DISPLACEMENT

WHEN SPACE ITSELF BREAKS DOWN

At the center of our galaxy lurks a monster. Sagittarius A* is a supermassive black hole containing the mass of four million suns, compressed into a region smaller than our solar system. It's so dense that if you could somehow transport a teaspoon of material from it to Earth, that teaspoon would weigh as much as Mount Everest.

Black holes represent the ultimate extreme in physics - regions where matter becomes so concentrated that space and time themselves break down. According to Einstein's general relativity, the gravitational field becomes so intense that not even light can escape. At the center lies a "singularity" - a point where the laws of physics as we know them simply stop working.

But what if black holes aren't mysterious breakdowns in the fabric of spacetime? What if they're simply regions where the substrate medium reaches its absolute limits? Akin to a bottle filled with water to the point that if you add any more it explodes.

The substrate model suggests that black holes are the universe's ultimate engineering failures - places where matter creates substrate displacement so extreme that the substrate itself breaks down, creating zones where the normal rules of substrate physics or any physics no longer apply. The same piecewise model governing subatomic particles governs the same switch at a macro level at the boundary of a black hole.

The Event Horizon: The Substrate Failure Boundary

The event horizon of a black hole - the boundary beyond which nothing can escape - isn't a physical surface. It's the boundary where the substrate medium fails completely governed by the piecewise structure, creating a void in the medium that makes physics possible.

Here's the crucial insight: light doesn't get "trapped" by black holes because of extreme gravitational pull. Light can't escape because light *cannot exist* in regions where the substrate medium has broken down. Since electromagnetic waves are pressure waves in the substrate medium, they simply cannot propagate through regions where no substrate exists.

Picture a sound wave trying to travel from air into a perfect vacuum. The sound doesn't get "trapped" at the boundary - it simply cannot continue because sound requires a medium to exist. At the vacuum boundary, the sound wave stops because there's no medium left to carry it.

Similarly, at a black hole's event horizon, light waves stop because there's no substrate medium left to carry them. The event horizon marks the boundary between regions where the substrate still functions (allowing electromagnetic waves to propagate) and regions where the substrate has failed completely (where electromagnetic waves cannot exist at all).

This is why nothing can escape from inside a black hole - not because they're being pulled back by extreme gravity, but because the very medium that allows forces, fields, and waves to exist has broken down. It's not that escape velocity exceeds light speed; it's that the substrate void prevents any information from propagating outward at all.

Beyond the Event Horizon: Substrate Voids and Compressed Boundaries

What happens inside a black hole, beyond the event horizon? According to general relativity, space and time become so distorted that normal descriptions break down. According to the substrate model, something more concrete and visualizable happens: the substrate medium gets completely expelled from the interior region.

The Complete Displacement

Imagine a swimming pool filled with water, but instead of dropping bowling balls into it, you begin pouring concrete into the deep end. As more concrete flows in, it doesn't just push the water aside temporarily - it **completely replaces** the water in that region. The water can't coexist with the concrete; it gets forced out entirely and can never flow back into the concrete-filled space.

Now imagine pouring so much concrete that it fills the entire deep end of the pool. The water isn't just displaced - it's **completely expelled** from that entire region. Where once there was water enabling fluid dynamics and wave propagation, now there's solid concrete where water-based physics simply cannot operate.

This is what happens inside a black hole. The extreme matter concentration creates displacement effects so intense that the substrate doesn't just get compressed or turbulent - it gets **completely pushed out**, creating a genuine void in the medium. The region inside the event horizon becomes like the concrete-filled section of the pool: a space where the substrate medium is entirely absent.

Why Physics Breaks Down

This explains why the normal laws of physics seem to break down inside black holes. It's not that the laws themselves fail - it's that the medium that makes those laws possible is completely absent. Just as you can't have waves or fluid dynamics in the concrete-filled section of the pool, you can't have gravitational, electromagnetic, or nuclear forces when there's no substrate medium to create pressure gradients and circulation patterns.

The forces don't become infinite or chaotic - they simply cease to exist because the medium that generates them is gone. Inside the event horizon, matter exists in a state that our substrate-based physics cannot describe, because substrate-based physics requires substrate to operate.

Where the Displaced Substrate Goes

But here's the crucial question: when you pour concrete into the pool, the water doesn't disappear - it gets pushed to other parts of the pool, raising the water level everywhere else. What happens to all the substrate that gets expelled from inside the black hole?

The answer reveals the true nature of the singularity.

The Singularity as Compressed Boundary

At the very center of a black hole lies the singularity - a region where general relativity predicts infinite density and infinite curvature of spacetime. Most physicists believe this represents a breakdown in our theories rather than a physical reality, but they haven't been able to explain what actually happens there.

The substrate model suggests a concrete answer: the singularity is not the central matter reaching infinite density. Instead, it's the **compressed boundary layer** where all the substrate displaced from the interior gets squeezed together.

Return to the pool analogy: when you fill the deep end with concrete, the displaced water doesn't vanish - it accumulates around the edges of the concrete, creating a higher water level at the boundary. The water gets compressed into a smaller volume around the concrete-filled region.

Similarly, when matter creates a black hole, it doesn't destroy the substrate - it pushes all of it to the central boundary, creating an incredibly dense substrate shell around the matter. All the substrate that once filled the entire interior region gets compressed into an almost infinitely thin layer surrounding the central mass.

The True Nature of Infinite Density

The "singularity" might actually be this compressed substrate boundary layer, containing all the displaced substrate from the entire black hole interior squeezed into an infinitesimally small shell. This explains why the singularity appears to have extreme properties - it's not the central matter that's infinite, but the compressed substrate boundary that contains the displaced medium from a region potentially light-years across, all crammed into a space smaller than an atom. Not infinite.

This resolves the infinite density paradox. The central matter doesn't become infinite - it exists in a substrate-free void. The apparent infinite density comes from the substrate boundary layer, where the displaced medium from the entire interior gets compressed to extreme density.

A Complete Picture

This gives us a complete, mechanically comprehensible picture of black hole structure:

Outside the event horizon: Normal substrate density supporting normal physics

At the event horizon: Critical boundary where substrate displacement becomes complete

Interior region: Genuine void with no substrate medium, where substrate physics cannot operate

Central boundary (singularity): Compressed shell containing all displaced substrate from the interior

Black holes aren't mysterious distortions of spacetime - they're regions where matter concentration has become so extreme that it completely displaces the substrate medium, creating voids surrounded by compressed substrate boundaries.

The event horizon isn't where spacetime curves infinitely - it's where the substrate medium gets completely expelled, creating a boundary between substrate-supported physics and substrate-free regions where our normal understanding of forces and fields simply doesn't apply.

This transforms black holes from incomprehensible mathematical abstractions into understandable mechanical systems: matter so dense it creates complete substrate displacement, resulting in interior voids surrounded by compressed substrate shells.

Hawking Radiation: Substrate Reconnection

One of the most remarkable predictions about black holes is Hawking radiation - the idea that black holes actually emit particles and slowly evaporate over time. This seems paradoxical: how can something that traps everything, allegedly including light, somehow emit particles?

Stephen Hawking's original explanation involved quantum fluctuations near the event horizon, but the substrate model suggests a more mechanical process: substrate reconnection.

When the substrate breaks down inside a black hole, it creates discontinuities in the medium - regions where the normally continuous substrate has gaps or tears. But nature abhors discontinuities. The substrate constantly tries to repair these tears through a process analogous to magnetic reconnection in plasma physics.

When torn magnetic field lines in a plasma reconnect, they release energy in the form of high-energy particles. Similarly, when torn substrate patterns around a black hole reconnect, they release energy in the form of particle-antiparticle pairs. One particle from each pair falls into the black hole, while the other escapes as Hawking radiation.

The black hole slowly evaporates not because of mysterious quantum effects, but because the substrate medium is constantly trying to repair the damage caused by extreme matter concentration. Each reconnection event extracts a tiny amount of mass-energy from the black hole and converts it into escaping particles.

Information Paradox: Substrate Memory

The "information paradox" asks what happens to information that falls into a black hole. According to quantum mechanics, information cannot be destroyed, but according to general relativity, anything that falls past the event horizon can never escape. When the black hole eventually evaporates through Hawking radiation, where does the information go?

The substrate model suggests that information falling into a black hole doesn't disappear - it gets encoded in the substrate reconnection patterns that generate Hawking radiation. The substrate medium has a kind of "memory" that preserves information about everything, including that which has fallen into the black hole.

Think of it like water flowing over a waterfall. The individual water molecules that go over the edge are lost to the pool below, but information about the flow pattern gets encoded in the spray, mist, and turbulence created by the falling water. An sufficiently sensitive observer could, in principle, reconstruct information about the original flow by analyzing these secondary effects.

Similarly, as matter falls into a black hole and disrupts the substrate medium, information about that matter gets encoded in the complex substrate reconnection patterns near the event horizon. The Hawking radiation carries this encoded information away from the black hole, preserving it even as the original compressed particles are stored into the singularity.

Primordial Black Holes: A Substrate Mechanics Paradox

For decades, physicists have theorized about primordial black holes - tiny black holes that supposedly formed in the early universe from extremely dense regions of matter. These microscopic black holes, some as small as atoms but carrying the mass of asteroids, have been proposed as candidates for dark matter and subjects for "substrate archaeology."

There's just one problem: despite extensive searching using increasingly sophisticated detection methods, no primordial black hole has ever been observed.

Substrate mechanics reveals why they can't exist.

In the substrate model, black holes form when matter concentration becomes so extreme that it completely displaces the substrate medium, creating genuine voids. But this displacement requires overcoming enormous substrate resistance - and that resistance becomes proportionally stronger at smaller scales.

Consider the physics: displacing substrate from even a microscopic volume requires tremendous energy density. At atomic scales, the substrate-to-matter ratio becomes so large that the available matter simply cannot generate enough pressure to push the substrate out completely. It's like trying to create a bubble in thick honey using only a grain of sand - there's insufficient force to displace the medium.

We see this same principle operating in stellar evolution. Not all stars become black holes when they die - only the most massive stars, those above roughly 25 solar masses, have enough matter density to overcome substrate resistance and create complete displacement. Smaller stars become white dwarfs (substrate compressed but intact) or neutron stars (substrate extremely compressed but not expelled), because they lack sufficient mass to push the substrate out entirely.

This creates a natural **substrate displacement hierarchy**: if even a 20-solar-mass star often can't generate enough pressure to displace substrate completely, then microscopic concentrations of primordial matter certainly can't. The substrate resistance that prevents most stellar deaths from creating black holes would be absolutely insurmountable for tiny primordial mass concentrations.

This suggests a fundamental **minimum mass threshold** for black hole formation. Below this threshold, matter lacks the density required to complete substrate displacement, making microscopic black holes physically impossible in substrate mechanics.

The continued failure to detect primordial black holes after decades of searching isn't a limitation of our instruments - it's confirmation that these objects don't exist because they can't exist. Substrate physics naturally explains their absence, while conventional black hole theory struggles to account for why something that should be abundant throughout the cosmos remains completely undetectable.

Rather than "substrate archaeology," studying the absence of primordial black holes becomes evidence for substrate mechanics itself. Their non-existence supports the prediction that black hole formation requires sufficient mass density to overcome substrate resistance - something that scales from stellar death outcomes down to microscopic impossibilities.

The universe isn't hiding tiny black holes from us. It's telling us that substrate displacement has fundamental limits, and we're finally learning to listen.

White Holes: Substrate Creation

If black holes represent regions where matter completely displaces the substrate medium, then the substrate model predicts the existence of their opposite: white holes, where the compressed substrate boundaries collapse and violently eject matter and energy back into the substrate medium.

While black holes have event horizons that nothing can escape from, white holes would have event horizons that nothing can enter. They would be regions where compressed substrate is releasing stored matter and energy so rapidly that particles flow outward faster than light, preventing anything from approaching the center.

This includes light itself. If nothing can penetrate the outward flow, then no light from our region of space can reach the white hole interior, and no information about what's happening inside can reach us. White holes would be completely invisible to external observers - we could never see them directly, no matter how powerful our telescopes.

This invisibility might explain why white holes have never been observed, despite being a natural prediction of physics. They're not absent from the universe - they're simply undetectable because their very nature prevents any observational evidence from reaching us.

The mechanism would be the reverse of black hole formation: when the compressed substrate boundary layer becomes unstable, it rapidly expands, converting its compressed substrate energy back into matter and radiation. This ejected matter then moves through the normal substrate medium, creating the outward flow that defines a white hole.

White holes might be so extraordinarily rare that they represent **cosmic regeneration events themselves** - the substrate boundary collapse that triggers universal rebirth. Rather than being scattered throughout the universe like black holes, white holes might be singular events that restart entire cosmic cycles.

When substrate energy accumulates across cosmic time and reaches critical density throughout the universe, the resulting white hole event wouldn't be a localized phenomenon - it would be **the Big Bang itself**. The compressed substrate boundary containing energy from countless black holes across cosmic history suddenly becomes unstable and releases everything at once.

This means we can account for white holes in our physics without expecting to observe them during normal cosmic evolution. They exist, but as **cycle tran-**

sition events rather than ongoing astronomical phenomena. The white hole that created our current universe allegedly happened 13.8 billion years ago, and the next one might not occur for another hundred billion years when substrate energy density again reaches critical levels.

We live in the aftermath of the last white hole event and in the buildup toward the next one.

Every black hole formation.

Every stellar explosion.

Every high-energy event.

All add to the substrate energy reservoir that will eventually trigger the next cosmic white hole - the next Big Bang.

This makes white holes both theoretically necessary and observationally absent: they're the rarest events in cosmic history, occurring only when an entire universe's worth of accumulated substrate energy releases simultaneously to begin the next cosmic cycle.

Wormholes: Substrate Shortcuts

Einstein's equations predict the theoretical possibility of wormholes - tunnels through spacetime that could connect distant regions of the universe. In the substrate model, wormholes would be regions where substrate breakdown creates "shortcuts" through the medium.

Imagine a river flowing around a large island. Normally, water has to follow the long path around the island to get from one side to the other. But if you could dig a tunnel through the island, water could flow directly from one side to the other, creating a shortcut that bypasses the normal flow pattern.

Similarly, if extreme matter concentrations could create "tunnels" in the substrate medium, these tunnels might provide shortcuts for substrate waves, effectively connecting distant regions of space through faster-than-light substrate wave propagation.

Such wormholes would allow faster-than-light travel for matter or energy through controlled substrate pressure manipulation similar to the Alcubierre warp drive concept.

Instead of warping spacetime itself, imagine creating artificial substrate pressure gradients that approach the black hole regime ($\varphi \rightarrow 0$). By generating regions of extremely low substrate density in front of a spacecraft and higher density behind it, you could create a "substrate wave" that carries the spacecraft along without the craft itself exceeding light speed relative to the local substrate.

The beauty of this approach is that it would also address the inertia problem. Since inertia emerges from substrate resistance to acceleration, reducing the substrate density in front of the spacecraft would automatically reduce inertial effects. The craft wouldn't need to "push" against substrate resistance during acceleration because the low-pressure zone ahead would eliminate that resistance.

This substrate-based warp drive would work by manipulating the same pressure gradients we see in black hole formation, but in a controlled, dynamic way that creates motion rather than gravitational collapse.

The Ultimate Test

Black holes might provide the ultimate test of the substrate model. If forces really emerge from substrate displacement mechanics, then the regions around black holes should show characteristic signatures as the substrate approaches its breaking point.

Current gravitational wave detectors are already observing black hole mergers - events where two regions of substrate breakdown collide and merge. The gravitational waves from these events might carry detailed information about substrate dynamics under the most extreme conditions possible.

Future space-based gravitational wave detectors might be sensitive enough to detect the subtle signatures of substrate reconnection, Hawking radiation, and other substrate breakdown effects. These observations could provide direct evidence for substrate physics operating at its absolute limits.

The Beauty of Breakdown

Perhaps the most remarkable aspect of understanding black holes through substrate mechanics is how it transforms them from mysterious cosmic monsters into comprehensible physical systems. Instead of regions where the laws of physics break down incomprehensibly, we have regions where a physical medium reaches its limits and fails in predictable ways.

Black holes become the universe's ultimate stress test - natural laboratories where we can study how the substrate medium behaves when pushed beyond its normal operating limits. They're not violations of physical law, but rather demonstrations of what happens when physical law reaches its boundaries.

The singularity at a black hole's center isn't a mathematical impossibility where physics stops working. It's simply the region where the substrate medium that makes physics possible is no longer present. The mystery isn't that physics breaks down there - the mystery is subatomic particles crushed together to form a void in the substrate.

Black holes reveal the substrate universe at its most extreme, showing us both the incredible resilience of the medium that makes physics possible and the dramatic consequences when that medium finally reaches its breaking point.

"Black holes ain't as black as they are painted. They are not the eternal prisons they were once thought."

Hawking

CHAPTER 14

THE COSMIC CRYSTAL

HOW SUBSTRATE GEOMETRY STORES UNIVERSAL INFORMATION

What if the universe has a perfect memory? What if every event, every pattern, every piece of information that has ever existed gets preserved in a cosmic storage system more sophisticated than any computer we could imagine?

The substrate doesn't just carry forces - it might be the universe's information storage medium, preserving knowledge through geometric patterns that tessellate perfectly throughout all of space.

The Perfect Information Geometry

Imagine spheres of substrate squeezed together under cosmic pressure, like soap bubbles in foam. Each sphere naturally wants to be round, but when pressed against its neighbors with no empty space between them, they deform into something beautiful: perfect hexagonal tessellations.

Look at a honeycomb and you're seeing this principle in action. Bees don't plan hexagons - they build circular cells that naturally deform into hexagonal perfection when pressed together. The substrate works the same way: spherical elements compressed into perfect tessellating geometry with no wasted space.

This isn't just efficient packing. It's **information architecture** formed by natural geometric principles.

When substrate elements contact each other, they form these same tessellating patterns throughout all of space. But unlike static crystal structures, substrate tessellations can store information through geometric deformations - subtle changes in angles, distances, and orientational relationships that encode vast amounts of data.

The Substrate Crystal Library

Think of the substrate as a three-dimensional crystal that fills all of space, but instead of storing just structural information like ordinary crystals, it stores **everything**:

The exact pattern of every particle interaction.

The precise details of every molecular structure that has ever existed.

The complete information content of every star, planet, and galaxy.

The neural patterns of every conscious thought that has ever occurred.

Each tessellation pattern becomes a storage cell in a cosmic library, with geometric relationships encoding information like data stored in crystalline memory.

How Information Gets Written

When high-energy events occur - stellar explosions, particle collisions, even complex biological processes - they create **pressure waves** in the substrate. These waves don't just dissipate randomly. They reorganize the local tessellation patterns, creating geometric records of the events.

A supernova explosion doesn't just release energy into space. It writes the **complete pattern** of that explosion into the substrate's geometric structure, preserving every detail in the altered tessellation around the event.

Your thoughts right now are creating subtle electrical patterns in your brain. These patterns generate tiny substrate pressure variations that modify local tessellation geometry. The substrate is literally recording your thoughts in its crystalline structure.

Staggering Storage Scale

To grasp the true magnitude of substrate information storage, consider this: every human who has ever lived - roughly 100 billion people - has unique DNA containing about 3 billion base pairs each. The complete genetic information for every human in history would require approximately 300 billion billion bits of data.

That enormous amount of information would fit comfortably on the tip of a sharpened pencil, using only a tiny fraction of the available storage space. We're talking about 99.999999999% of the space still remaining empty.

And that's just at the scale of pencil tips. The substrate tessellation extends through every cubic inch of space throughout the entire universe.

With substrate elements potentially at sub-atomic scales, each tessellation junction could store information through:

Precise angles between connecting elements.

Exact distances between neighboring points.

Rotational orientations of local patterns.

Harmonic frequency relationships.

Phase relationships between adjacent regions.

With substrate elements potentially smaller than atomic scales by a factor of 100, the information density could exceed anything we can imagine - perhaps storing the complete state of entire solar systems in volumes smaller than a grain of sand.

The Heat Death Solution Through Information

This geometric information storage completely transforms the heat death problem. Heat death assumes information gets lost to thermal randomization - that the precise arrangements that create complexity get scrambled into useless chaos.

But substrate tessellation patterns preserve information even when local matter becomes thermally randomized. The geometric record remains intact in the substrate crystal structure, waiting to be accessed.

When substrate energy density reaches critical levels across cosmic time, the information doesn't trigger just energy release - it enables **pattern reconstruction**. The stored geometric information becomes the template for regenerating all the complex structures that existed in previous cosmic cycles.

Cosmic Information Inheritance

Each universe inherits the complete information library of all previous universes. Stars form in patterns influenced by geometric information from stellar formations billions of cycles ago. Planetary systems organize according to templates preserved in substrate tessellations. Even biological evolution might follow pathways encoded in the cosmic crystal.

We're not starting from scratch in a random universe. We're inheriting the accumulated wisdom and organizational patterns of infinite previous cosmic cycles, all preserved in the geometric structure of the substrate itself.

The Living Cosmic Memory

The substrate becomes a living memory system - constantly recording new information while preserving ancient patterns. Every moment adds to the cosmic library while maintaining access to all previous knowledge.

This means consciousness and intelligence aren't accidents in a dying universe. They're **information processing systems** that naturally emerge from substrate tessellation patterns, accessing and contributing to the cosmic information storage that spans all of time.

The Tessellation Network

The beautiful part is how information connects across vast distances. Tessellation patterns that touch create information networks spanning cosmic scales. A thought in your mind creates geometric patterns that connect through substrate tessellations to similar patterns throughout the universe.

Consciousness might be networked through the substrate crystal structure - not just locally in individual brains, but cosmically through the geometric information patterns that tessellate throughout all of space.

The Eternal Library

In the substrate universe, nothing is ever truly lost. Every pattern, every structure, every thought gets preserved in the cosmic crystal. Heat death becomes impossible because information, once encoded in tessellation geometry, remains accessible for reconstruction across infinite time.

The universe isn't running down toward chaos. It's building an ever-more sophisticated information storage system, accumulating knowledge and complexity across eternal cycles of cosmic death and rebirth.

We're not temporary accidents in a doomed cosmos. We're information processors participating in cosmic intelligence - reading from and writing to a universe-spanning crystalline memory that preserves and builds upon its achievements across infinite time.

The substrate crystal remembers everything. And through our consciousness, we help it learn.

"This technology can secure the last evidence of our civilization: all we've learned will not be forgotten."

Kazansky

CHAPTER 15

THE ETERNAL COSMOS

WHEN HEAT DEATH BECOMES REBIRTH

One of physics's most haunting predictions is the heat death of the universe - the idea that all energy will eventually spread out uniformly, leaving a cold, lifeless cosmos where nothing interesting can ever happen again. It's a vision that transforms the universe from a magnificent story into a tragic tale with an inevitably bleak ending.

But what if heat death isn't inevitable? What if the substrate changes everything about how we understand cosmic destiny?

The Heat Death Problem

According to classical thermodynamics, the universe is like a giant engine running down. Every process that generates useful energy - stars burning, planets forming, life evolving - also creates waste heat that spreads out and can never be recovered. The second law of thermodynamics tells us that entropy always increases, meaning the universe becomes progressively more disordered until it reaches a state of maximum entropy: perfect thermal equilibrium at a temperature just above absolute zero.

In this heat death scenario, all stars eventually burn out, all black holes evaporate, and the universe becomes an endless expanse of cold, sparse particles drifting apart forever. No work can be done, no information can be processed, no life can exist. The universe becomes a cosmic graveyard where nothing will ever happen again.

This prediction has troubled physicists and philosophers for over a century. It suggests that everything we consider meaningful - consciousness, creativity, love, discovery - is just a temporary accident in an ultimately purposeless cosmos.

The Substrate Revolution

But substrate theory suggests a completely different ending to the cosmic story. Instead of heat death, we might have eternal cycling - a universe that regenerates itself through substrate energy storage and release.

Here's how it works: When high-energy events occur - supernovas, black hole formations, particle collisions - the substrate doesn't just passively transmit the effects. It absorbs and stores a fraction of that energy in highly organized wave patterns.

Think of it like this: imagine you're watching waves crash on a beach. Most of the wave energy dissipates as random thermal motion in the sand and water. But suppose the beach could somehow capture and store some of that wave energy in organized underground patterns - structured energy reservoirs that grow larger with each crashing wave.

The substrate might work similarly. Each cosmic explosion, each stellar death, each high-energy event adds organized energy to vast substrate reservoirs. Unlike thermal energy, which is chaotic and useless for doing work, substrate energy remains organized and available.

The Mathematics of Cosmic Immortality

The mathematics are startling. In classical thermodynamics, every process increases total entropy, pushing the universe toward maximum disorder. But when the substrate stores energy in organized patterns, something remarkable happens: the universe can actually decrease in total entropy.

Consider a supernova explosion. Classically, this creates an enormous increase in entropy as organized stellar matter becomes random thermal energy radiating into space. But if the substrate absorbs even a small fraction of this energy in organized wave patterns, the entropy decrease from substrate organization can actually exceed the entropy increase from thermal randomization.

The key is the temperature ratio. Thermal energy exists at temperatures of millions of degrees, while substrate energy is stored in organized patterns at much lower effective temperatures. When energy transfers from hot, chaotic thermal systems to cool, organized substrate patterns, total entropy can decrease even while energy is conserved.

This is like watching a pot of boiling water suddenly become more organized rather than less - impossible in normal thermodynamics, but theoretically possible when you include substrate energy storage.

Cosmic Cycling: From Death to Rebirth

As the universe ages and more high-energy events occur, substrate energy reservoirs grow larger and more organized. Eventually, they reach a critical density where the substrate itself undergoes a phase transition - like water suddenly turning to steam, but in reverse.

When substrate energy density reaches this critical point, the organized substrate patterns become unstable and release their stored energy all at once. This creates a cosmic regeneration event - essentially a new Big Bang powered by the accumulated substrate energy from the previous cosmic cycle.

The universe doesn't die in heat death. Instead, it regenerates through substrate energy release, creating new matter, new energy, new possibilities for complexity and life. As predicted by the ultra rare white hole phenomenon.

Each cosmic cycle lasts perhaps a hundred billion years or more - long enough for stars to form, planets to develop, life to evolve, and civilizations to arise. But instead of ending in cold death, each cycle concludes with rebirth as substrate energy triggers the next cosmic beginning.

Information Preservation Across Cycles

Perhaps most remarkably, information might survive across cosmic cycles. Just as substrate stores energy in organized patterns, it might also preserve information in these same patterns. The substrate becomes a cosmic memory system, maintaining records of previous cycles that influence the structure of future ones.

This means each cosmic cycle builds on the achievements of previous cycles. Information accumulates across infinite regenerations, creating a universe that becomes progressively more sophisticated rather than progressively more random.

Life and consciousness aren't accidents in a dying cosmos - they're part of an eternal cosmic story that grows richer with each cycle. Every discovery we make, every work of art we create, every moment of beauty or understanding might be preserved in substrate memory patterns that influence future cosmic cycles.

The Molecular Connection

This cosmic story connects directly to the molecular frequencies we observe in everyday chemistry. When molecules vibrate at characteristic frequencies, they might be tapping into the same substrate wave patterns that will eventually power cosmic regeneration.

The hydrogen molecule vibrating at 4,161 wavenumbers per centimeter might be participating in substrate dynamics that scale up through nuclear physics (at 2.4 GeV frequencies) all the way to cosmic cycling timescales. The same substrate that holds atoms together might be accumulating the energy that will birth the next universe.

This suggests a profound unity: the forces that create chemical bonds operate through the same substrate mechanics that ensure cosmic immortality. Every molecule is both a local chemical phenomenon and a participant in eternal cosmic cycling.

From Tragedy to Symphony

This transforms our understanding of cosmic destiny from tragedy to symphony. Instead of a universe that starts with a bang and ends with a whimper, we have a cosmos that cycles eternally between organization and reorganization, death and rebirth, ending and beginning.

The heat death of the universe becomes just the quiet movement between cosmic symphonies - a brief rest before the substrate releases its accumulated energy in a new crescendo of matter, energy, and possibility.

Consciousness and intelligence aren't meaningless accidents in a doomed cosmos. They're part of an eternal creative process that preserves and builds upon its achievements across infinite cycles. We're not temporary anomalies in a dying universe - we're participants in cosmic immortality.

The Eternal Questions

If substrate cycling proves correct, it raises profound questions about continuity and identity across cosmic cycles. Does information preservation mean that consciousness itself might continue in some form? Could civilizations that achieve deep understanding of substrate mechanics influence the parameters of future cycles?

These questions push us beyond physics into philosophy and meaning. But they're questions asked within a framework of cosmic optimism rather than cosmic pessimism. Instead of asking how to find meaning in a doomed universe, we ask how to participate meaningfully in eternal cosmic creativity.

The substrate universe doesn't just make physics comprehensible - it makes existence meaningful across infinite time. Heat death becomes heat regeneration. Cosmic ending becomes cosmic continuation. The universe transforms from a tragedy with an inevitable sad ending to an eternal story that never truly ends, but only pauses between movements. In the substrate cosmos, the music never stops playing. It only gets more beautiful with each cosmic symphony.

"Although mechanical energy is indestructible, there is a universal tendency to its dissipation, which produces throughout the system a gradual augmentation and diffusion of heat, cessation of motion and exhaustion of the potential energy of the material Universe."

Kelvin

CHAPTER 16

ENGINEERING THE INVISIBLE

BUILDING TOMORROW'S TECHNOLOGY FROM TODAY'S DISCOVERY

Imagine walking into your garage and seeing a car floating three feet off the ground, perfectly stable, consuming no energy to maintain its position. Imagine picking up your phone and having an instant conversation with someone on Mars, with no delay despite the millions of miles between you. Imagine powering your entire city with a clean nuclear reactor the size of a refrigerator, producing no radioactive waste.

These aren't science fiction fantasies. If the substrate model is correct, they're engineering challenges - difficult ones, but no more impossible than flight seemed to the Wright brothers or global wireless communication seemed to Marconi.

Every breakthrough in physics has eventually led to transformative technologies. Understanding electricity led to motors, generators, and computers. Understanding electromagnetic waves led to radio, television, and wireless communication. Understanding nuclear physics led to nuclear power and medical imaging.

Understanding substrate mechanics might lead to the most transformative technologies in human history: direct control over gravity, instantaneous communication across any distance, and clean nuclear energy systems that work with the substrate rather than against it.

The Substrate Engineering Revolution

For the first time in physics, we might have a unified understanding of how all forces actually work - not through mysterious fields acting at a distance, but through pressure gradients and circulation patterns in a physical medium. This opens up an entirely new field: substrate engineering.

Just as we learned to control electromagnetic fields once we understood their mechanism, we might learn to control gravitational fields, nuclear forces, and even the propagation of information itself once we master substrate manipulation.

Think of substrate engineering as the ultimate form of environmental control. Instead of just controlling air pressure, water flow, and temperature in our environment, we'd be controlling the pressure gradients and circulation patterns in the very medium that creates all physical forces.

The technologies that emerge from this could be as revolutionary as the difference between a stone age cave and a modern city.

Gravity Control: The Holy Grail

Gravity control has been humanity's dream since we first watched birds fly. If gravity really operates through pressure gradients in the substrate medium, then controlling gravity becomes a matter of learning to manipulate those pressure gradients artificially.

Imagine being able to create artificial substrate displacement zones around a spacecraft. The higher substrate pressure at the boundaries of these zones would squeeze the spacecraft toward the displacement regions, creating controlled acceleration in any direction you choose. This works by manipulating the pressure differential, not by creating 'attraction' to low pressure. You could create "gravitational slopes" that pull your vehicle upward, forward, or in any direction you desire.

Personal Gravity Control At the smallest scale, personal gravity control devices might work like substrate pressure suits. By creating carefully controlled substrate displacement zones around your body, you could manipulate the pressure differentials that create your apparent weight. Higher boundary pressure would squeeze you toward these artificial displacement zones, enabling controlled flight through pressure gradient manipulation. Walking on the ceiling would be as easy as creating appropriate pressure gradients above your head.

Spacecraft Propulsion For spacecraft, substrate propulsion could revolutionize space travel. Instead of carrying enormous amounts of fuel to push against, you'd create substrate pressure gradients that pull you toward your destination. A Mars mission could work by creating an artificial low-pressure zone near Mars, allowing the natural substrate pressure around Earth to push the spacecraft across the solar system.

The energy requirements might be surprisingly modest. You wouldn't be fighting against gravity - you'd be redirecting the substrate pressure gradients that create gravity. It's like redirecting a river rather than trying to stop it.

Orbital Mechanics Revolution Satellites could maintain perfect orbits without fuel, by using small substrate pressure adjustments to correct their paths. Space elevators might become practical if you could reduce the substrate pressure gradients along their length, effectively making the cable "lighter" without changing its actual mass.

Substrate Communication: Instant Connection

If substrate waves really propagate at billions of times the speed of light, then substrate-based communication could make distance irrelevant. Instead of waiting minutes for radio signals to travel between planets, substrate communication could be genuinely instantaneous across any distance.

The Substrate Internet Imagine a communication network based on substrate wave patterns rather than electromagnetic signals. Your phone wouldn't send radio waves to cell towers - it would create specific patterns in the local substrate that instantly propagate to substrate receivers anywhere in the universe.

The "substrate internet" would make the current internet seem slow and limited. Every device could be instantly connected to every other device, regardless of distance. Lag would become a thing of the past. Real-time communication between Earth and Mars colonies would be as easy as talking to someone in the next room.

Navigation Revolution Current GPS systems rely on timing signals from satellites, but substrate positioning could be both faster and more accurate. Substrate wave patterns could provide instant, three-dimensional position information anywhere in the solar system, unaffected by obstacles, weather, or electromagnetic interference.

A spacecraft near Jupiter could know its exact position relative to Earth instantly, without waiting for radio signals to cross the solar system. Autonomous vehicles could navigate with perfect precision using substrate positioning that works underground, underwater, or anywhere else electromagnetic signals can't reach.

Search and Rescue Substrate communication could revolutionize emergency response. Survivors buried under avalanches or trapped in collapsed buildings could communicate instantly with rescue teams, regardless of how much rock, ice, or debris blocks electromagnetic signals. Submarine crews could maintain constant contact with the surface without having to deploy antenna buoys.

Nuclear Void Engineering: Clean Energy Revolution

Understanding nuclear forces as extreme substrate pressure effects could lead to completely new approaches to nuclear energy - approaches that work with substrate mechanics rather than fighting against them.

Controlled Nuclear Voids Instead of trying to smash nuclei together with brute force (fusion) or break them apart with neutron bombardment (fission), substrate engineering might allow direct manipulation of the nuclear pressure wells that hold atoms together.

Imagine being able to create artificial nuclear voids - regions of controlled substrate displacement that mimic the pressure wells inside atomic nuclei. By manipulating these artificial voids, you might be able to:

Guide fusion reactions with perfect precision, making fusion power practical and safe.

Neutralize radioactive waste by manipulating the substrate pressure wells around unstable nuclei.

Create new elements with desired properties by engineering specific nuclear pressure configurations.

Extract nuclear energy on demand without the massive infrastructure of current nuclear plants.

The Substrate Reactor A substrate-based nuclear reactor might be surprisingly compact and safe. Instead of containing a critical mass of fissile material in massive shielded buildings, it might use substrate manipulation to create controlled nuclear void interactions in tiny amounts of fuel.

Such a reactor could be inherently safe because it would only work when the substrate manipulation systems are actively maintained. Any malfunction would automatically shut down the nuclear reactions by eliminating the artificial substrate conditions required to sustain them.

The waste products might be minimal because substrate engineering could guide nuclear reactions toward stable end products rather than radioactive byproducts. Instead of creating long-lived radioactive waste, the reactor could produce only stable elements that pose no environmental hazard.

Substrate Manufacturing: Building with Pressure

If chemical bonds are really overlapping substrate pressure wells, then substrate engineering could revolutionize manufacturing by allowing direct manipulation of atomic and molecular structures.

Programmable Matter Imagine materials whose atomic structure could be reconfigured on command by adjusting the substrate pressure patterns that hold them together. A single block of "programmable matter" could be reconfigured into different materials with different properties - now diamond-hard for cutting tools, now flexible like rubber for gaskets, now conductive like copper for wiring.

Molecular Assembly Manufacturing could become as precise as assembling structures with building blocks. By creating specific substrate pressure well patterns, you could guide atoms and molecules into any desired configuration. Building a computer chip could be as simple as creating the right substrate pressure template and letting atoms naturally settle into the required pattern.

Self-Repairing Materials Materials could be designed with substrate pressure patterns that automatically correct defects. Scratches, cracks, or other damage could trigger substrate pressure adjustments that guide atoms back into their proper positions, creating materials that literally heal themselves.

Medical Applications: Healing with Substrate

Understanding biological processes as substrate pressure phenomena could revolutionize medicine by allowing direct manipulation of the substrate patterns that maintain cellular structure and function.

Precision Surgery Instead of cutting tissue with physical tools, substrate manipulation could allow surgeons to precisely adjust the pressure wells that hold cellular structures together. Removing tumors could be as simple as disrupting the substrate patterns that maintain cancer cell integrity while leaving healthy cells untouched.

Regenerative Medicine If biological development is guided by substrate pressure templates, then medical treatment could involve correcting or restoring the substrate patterns that guide tissue growth and repair. Regrowing lost limbs or organs might become possible by recreating the substrate pressure patterns that originally guided their development.

Neural Interface Technology The brain's electrical activity creates patterns in the substrate that might be detectable and even controllable with sufficient precision. Direct neural interfaces based on substrate manipulation could allow seamless connection between consciousness and artificial systems without the need for invasive implants.

The Challenges Ahead

Of course, substrate engineering faces enormous technical challenges. We're talking about manipulating a medium so subtle that we've barely begun to detect its effects, much less control them.

Detection Before Control Before we can engineer the substrate, we need to develop reliable methods for detecting and measuring substrate properties. This might require entirely new types of instruments - substrate sensors that can detect pressure gradients, circulation patterns, and wave propagation in a medium more subtle than anything we've previously studied.

Energy Requirements Some substrate engineering applications might require enormous amounts of energy to create the necessary pressure gradients or circulation patterns. The energy costs of gravity control or large-scale substrate manipulation might initially make these technologies impractical except for specialized applications.

Safety Considerations Manipulating the medium that creates all physical forces could have unexpected consequences. Substrate engineering experiments would need to be conducted with extreme caution to avoid creating uncontrolled substrate effects that could affect surrounding areas.

Material Science We might need entirely new materials designed to interact with substrate patterns in controlled ways. Current materials have properties determined by their electromagnetic interactions, but substrate engineering might require materials whose properties are determined by their substrate displacement and circulation patterns.

The Development Timeline

The progression toward practical substrate engineering might follow a predictable pattern, similar to the development of other technologies:

Phase 1: Basic Detection (Next 10-20 years) Development of sensitive instruments capable of detecting substrate pressure gradients, circulation patterns, and wave propagation. Confirmation of substrate wave speeds and basic substrate properties through precision measurements.

Phase 2: Controlled Effects (20-50 years) Creation of small-scale, controlled substrate effects in laboratory conditions. Demonstration of basic substrate pressure gradient generation and circulation pattern control. Development of substrate-based sensors and measurement devices.

Phase 3: Practical Applications (50-100 years) First practical substrate engineering devices for specialized applications. Substrate-based communication systems for space exploration. Small-scale gravity control for research and industrial applications. Substrate-enhanced nuclear energy systems.

Phase 4: Widespread Adoption (100+ years) Integration of substrate engineering into everyday technology. Personal gravity control devices. Substrate-based transportation systems. Programmable matter manufacturing. Medical applications of substrate manipulation.

The Investment Opportunity

The potential economic impact of substrate engineering could dwarf any previous technological revolution. The companies and nations that master substrate manipulation first could dominate industries worth trillions of dollars:

Transportation systems that don't require fuel.

Communication networks that work across interplanetary distances.

Manufacturing processes that can create any material on demand.

Energy systems that are both clean and unlimited.

Medical treatments that work at the cellular level with perfect precision.

The research and development costs would be enormous, but so would the potential returns. Substrate engineering could create entirely new industries while revolutionizing existing ones.

A Word of Caution

It's important to remember that substrate engineering remains highly speculative. The substrate model might be wrong, or the technological applications might prove impossible despite theoretical possibilities. Even if the basic theory is correct, the engineering challenges could be far more difficult than we imagine.

History is littered with technologies that seemed theoretically possible but proved practically impossible, at least with available resources and knowledge. Substrate engineering might fall into this category, remaining forever tantalizingly out of reach. But history is also full of technologies that seemed impossible until someone figured out how to make them work. Powered flight, wireless communication, nuclear energy, and quantum computers all seemed like fantasy until they became reality.

The Ultimate Technology

If substrate engineering becomes practical, it could represent the ultimate technology - direct control over the medium that makes all physics possible. Instead of working around the limitations imposed by gravity, electromagnetic fields, and nuclear forces, we could control those forces directly.

We could become engineers of reality itself, shaping the substrate pressure gradients and circulation patterns that create the world we experience. The boundary between the possible and impossible would shift dramatically in favor of human capability and imagination.

The invisible substrate that fills all space could become humanity's most powerful tool - and our gateway to capabilities that seem magical today but might someday be as commonplace as flipping a light switch.

"The present is theirs; the future, for which I really worked, is mine."

Tesla

CHAPTER 17

REWRITING THE TEXTBOOKS

HOW SCIENCE WILL CHANGE FOREVER

Picture a physics classroom in the not-too-distant future. A student raises her hand and asks, "Professor, why do opposite charges attract each other?"

Instead of launching into abstract mathematics about field theory and force vectors, the professor walks to a large tank of water at the front of the classroom. She drops two cork balls into the water, each with small motors that spin them in opposite directions, creating tiny whirlpools around each cork.

"Watch what happens when the whirlpools meet," she says.

The students observe as the opposing circulation patterns mesh together like gears, drawing the cork balls toward each other. The professor then reverses one motor, so both corks create whirlpools spinning in the same direction. Now the whirlpools interfere with each other, pushing the corks apart.

"Electric charge works exactly the same way," the professor explains, "except it's happening in the substrate medium that fills all space. Positive charges create circulation in one direction, negative charges in the opposite direction. When opposite circulations meet, they mesh together and attract. When similar circulations meet, they interfere and repel."

The student nods, her face lighting up with understanding. "So it's just like gears in a machine?"

"Exactly like gears in a machine. Physics is mechanical again."

The End of Mysterious Physics

For over a century, physics education has required students to abandon their intuitive understanding of how the world works. They've been told to accept that:

Particles can be in multiple places simultaneously.

Forces act mysteriously across empty space.

The quantum world operates by rules completely different from everyday experience.

The more you understand physics, the weirder it gets.

Students have been forced to memorize mathematical formulas and accept counterintuitive concepts without any mechanical understanding of what's actually happening. Many potentially brilliant physicists have been lost along the way, unable to make peace with a subject that seems to defy common sense.

The substrate model could change all of this. Instead of teaching physics as a collection of mysterious phenomena, we could teach it as mechanical engineering applied to the most subtle medium in the universe.

Physics Education Revolution

From Abstract to Intuitive

Current physics education moves from the familiar (classical mechanics) to the increasingly abstract (quantum field theory, general relativity, particle physics). Students start with concepts they can visualize and gradually learn to think in mathematical abstractions they can never truly picture.

Substrate-based physics education could work the opposite way. Students would start by learning about wave mechanics in media they can see and touch - water waves, sound waves, waves on strings. Then they'd learn that all of physics operates by the same wave mechanics principles, just in a medium they can't directly perceive.

Instead of learning separate, mysterious rules for different branches of physics, students would learn one set of mechanical principles that applies everywhere. Gravity, electromagnetism, nuclear forces, and quantum mechanics would all be applications of the same substrate wave mechanics.

Hands-On Understanding

Physics demonstrations could become truly meaningful again. Every wave tank demonstration, every oscillating spring, every fluid dynamics experiment would be directly applicable to understanding fundamental physics. Students could literally see how forces work by watching analogous processes in visible media.

A vibrating drumhead could demonstrate both sound waves and quantum wave functions. A whirlpool in a bathtub could illustrate both fluid dynamics and electromagnetic field circulation. A pressure wave in water could represent both sound propagation and gravitational wave transmission.

Visual Learning

Students who struggle with mathematical abstractions could understand physics through visual and mechanical intuition. Instead of memorizing that "the wave function collapses upon measurement," they could see how measurement interactions inevitably disturb delicate wave patterns in any medium.

Complex phenomena could be understood through simple analogies that are mechanically accurate rather than just metaphorically helpful. Wave-particle duality wouldn't be a mysterious concept - it would be as obvious as recognizing that a wave pulse in water looks localized (particle-like) but is actually a traveling pattern in the medium (wave-like).

The New Physics Curriculum

Foundations: Medium Mechanics Students would start by thoroughly understanding how waves and pressure systems work in familiar media - air, water, elastic materials. They'd learn about pressure gradients, circulation patterns, wave interference, and medium properties.

Universal Principles: Substrate Mechanics Next, students would learn that space itself is filled with a medium (the substrate) that responds to matter through displacement, creating pressure gradients and circulation patterns. All forces would be understood as mechanical effects in this medium.

Applications: Force Systems Different forces would be taught as different aspects of substrate mechanics:

Gravity as gentle, long-range pressure gradients.

Electromagnetism as circulation patterns and pressure waves.

Nuclear forces as extreme pressure in compressed substrate.

Quantum mechanics as wave mechanics in the substrate.

Advanced Topics: Substrate Engineering Advanced students would learn to calculate and predict substrate effects, design experiments to detect substrate properties, and explore technological applications of substrate manipulation.

Training Substrate Engineers

The next generation of physicists wouldn't just be theoretical researchers or experimental investigators - they'd be substrate engineers, trained to understand and manipulate the medium that makes all physics possible.

Interdisciplinary Training Substrate engineering would require knowledge from multiple fields:

Fluid dynamics for understanding substrate flow and pressure systems.

Wave mechanics for analyzing substrate oscillations and propagation.

Materials science for creating substances that interact with substrate in controlled ways.

Electrical engineering for building substrate detection and manipulation devices.

Computer science for modeling complex substrate interactions

Practical Skills Students would learn practical skills for working with substrate effects:

Designing and building substrate wave detectors.

Creating controlled substrate pressure gradients.

Measuring substrate circulation patterns.

Calculating substrate displacement effects for different mass configurations.

Engineering substrate-based devices and systems.

Research Focus Research priorities would shift toward understanding substrate properties and developing substrate manipulation techniques:

Mapping substrate density and elasticity across different environments.

Developing more sensitive substrate detection methods.

Creating artificial substrate pressure gradients and circulation patterns.

Investigating substrate wave propagation and interference.

Exploring biological interactions with substrate fields.

The Democratization of Physics

One of the most exciting prospects of substrate-based physics education is how it could make physics accessible to far more students. Currently, physics has a reputation as one of the most difficult subjects, requiring exceptional mathematical ability and a willingness to accept counterintuitive concepts.

Substrate physics could be as intuitive as engineering or medicine - fields that require precision and knowledge but don't ask students to abandon their mechanical understanding of how things work.

Visual and Kinesthetic Learners Students who learn best through visualization and hands-on experience could excel in substrate physics. Understanding forces through pressure gradients and circulation patterns appeals to spatial and mechanical intelligence rather than just mathematical abstraction.

Engineering Mindset Students with practical, engineering-oriented minds - who may have been put off by the abstract nature of current physics - might find substrate physics naturally appealing. It's applied mechanics on the grandest possible scale.

Diverse Applications Substrate physics would connect naturally to many fields:

Biology (how living systems interact with substrate fields).

Medicine (substrate-based diagnostic and treatment methods).

Environmental science (substrate effects in atmospheric and oceanic systems).

Computer science (substrate-based information processing).

Materials science (designing materials with specific substrate properties).

Research Methodology Revolution

The substrate model would also transform how physics research is conducted. Instead of increasingly abstract theoretical work divorced from mechanical understanding, research would focus on understanding and manipulating a physical medium.

Experimental Focus Research would become more experimentally oriented, focusing on:

Detecting and measuring substrate properties.

Creating controlled substrate effects.

Testing substrate-based technological applications.

Mapping substrate variations in different environments.

Theoretical Work Even theoretical research would be grounded in mechanical principles. Instead of developing increasingly abstract mathematical frameworks, theorists would work on:

Modeling substrate fluid dynamics in complex systems.

Calculating substrate wave propagation in various configurations.

Predicting substrate responses to different mass and energy distributions.

Designing optimal substrate manipulation strategies.

Collaborative Science Substrate research would naturally encourage collaboration between physicists, engineers, materials scientists, and biologists. Understanding a universal medium that affects all physical processes would require expertise from many fields.

The End of Physics Fragmentation

Current physics is divided into numerous specialties that seem almost unrelated:

Particle physics (studying fundamental particles and forces).

Condensed matter physics (studying materials and their properties).

Astrophysics (studying cosmic phenomena).

Quantum mechanics (studying atomic and subatomic behavior).

General relativity (studying gravity and spacetime).

Students often specialize in one area and lose sight of the connections between different branches of physics. The substrate model would reunify physics by showing that all these specialties are studying different aspects of the same underlying medium.

Unified Understanding A substrate physicist would understand that:

Particle physics studies extreme substrate displacement effects.

Condensed matter physics studies substrate interactions in complex systems.

Astrophysics studies substrate effects on cosmic scales.

Quantum mechanics studies substrate wave mechanics on small scales.

General relativity describes the geometric effects of substrate displacement.

Cross-Scale Insights Understanding gained in one area would directly apply to others. Techniques for studying substrate effects in atomic nuclei could inform research on neutron stars. Methods for detecting substrate waves in laboratory experiments could advance gravitational wave astronomy.

Philosophical Implications

The substrate model would also have profound philosophical implications for how we think about the nature of reality and scientific knowledge.

The Return of Mechanical Philosophy

For the first time since the early days of physics, we could have a completely mechanical understanding of natural phenomena. Instead of accepting mysterious "action at a distance" or incomprehensible quantum effects, everything would be explained through contact interactions in a physical medium.

This would restore the philosophical satisfaction that comes from truly understanding how things work, rather than just being able to calculate their effects.

The Unity of Science The substrate model suggests that all of science might ultimately be unified through understanding different aspects of substrate mechanics:

Chemistry: How substrate pressure wells create atomic and molecular structures.

Biology: How living systems organize and maintain substrate patterns.

Geology: How substrate effects influence planetary formation and dynamics.

Psychology: How consciousness might emerge from substrate pattern organization in neural systems.

Scientific Method Enhancement The substrate model would encourage a return to mechanical understanding as a goal of scientific research.

The Challenge of Transition

Of course, transitioning to substrate-based physics education would face significant challenges:

Faculty Training Current physics professors would need extensive retraining to teach substrate-based approaches. Many concepts they've taught for decades would need to be reconceptualized in terms of substrate mechanics.

Textbook Revolution Every physics textbook would need to be completely rewritten, starting from substrate principles rather than traditional field-based approaches. This would be an enormous undertaking requiring coordination among educators worldwide.

Institutional Resistance Universities and professional physics organizations might resist such fundamental changes to established curricula and research directions. The transition could take decades to complete fully.

Validation Requirements The substrate model would need extensive experimental validation before it could be adopted as the foundation for physics education. Students can't be taught speculative theories as established facts.

The Future

Despite these challenges, the potential benefits of substrate-based physics education are enormous. Imagine a generation of physicists who understand all of physics through mechanical principles, who can visualize and intuitively grasp phenomena from quantum mechanics to cosmology, and who are trained to engineer solutions using the most fundamental medium in the universe.

These physicists wouldn't see nature as mysterious and incomprehensible - they'd see it as the most elegant machine ever constructed, operating through principles as clear and logical as any human engineering project.

Physics would become not just the study of natural phenomena, but the engineering discipline for reality itself. And students would fall in love with physics not despite its complexity, but because of its underlying simplicity and beauty.

The textbooks of the future might begin not with mysterious forces and abstract mathematics, but with a simple statement: "Physics is the study of how pressure and flow in an invisible medium create everything we observe in the universe. Let's learn how this elegant machine works."

"Science cannot solve the ultimate mystery of nature. And that is because, in the last analysis, we ourselves are a part of the mystery that we are trying to solve."

Planck

CHAPTER 18

THE UNIVERSE REVEALED

WHAT THIS MEANS FOR EVERYTHING

As we reach the end of our journey through the substrate universe, it's important to pause and reflect on what we've discovered - and what we haven't. The substrate model doesn't replace the magnificent edifice of modern physics that took centuries to build. Einstein's equations still predict the motion of planets with extraordinary precision. Quantum mechanics still describes atomic behavior with incredible accuracy. The Standard Model still successfully predicts the results of particle accelerator experiments.

What the substrate model potentially provides is something different and complementary: a mechanical explanation for why these mathematical frameworks work so beautifully. It's not about overthrowing physics - it's about understanding physics more deeply.

Think of it this way: Newton's laws perfectly describe how objects move, but they don't explain why objects have inertia in the first place. Einstein's field equations precisely predict gravitational effects, but they don't explain what gravity actually is mechanically. Quantum mechanics accurately calculates atomic properties, but it doesn't explain why particles exhibit wave-like behavior.

The substrate model doesn't challenge these successful theories - it potentially explains the mechanical reality underlying their mathematical success.

The Complementary Revolution

Every great advance in physics has built upon rather than demolished what came before. When Einstein developed relativity, he didn't prove Newton wrong - he showed that Newton's laws are perfectly accurate approximations that work beautifully in everyday situations. Newton's equations are still used to land spacecraft on Mars and design bridges.

Similarly, if the substrate model proves correct, it wouldn't invalidate modern physics - it would provide the mechanical foundation that explains why modern physics works so well.

Einstein's Spacetime and Substrate Displacement Einstein showed us that massive objects curve spacetime, creating what we experience as gravitational effects. The substrate model suggests why this curvature occurs: massive objects displace the substrate medium, creating pressure gradients that manifest as apparent spacetime curvature.

Einstein's equations would remain perfectly valid for calculating gravitational effects. But we'd finally understand the mechanical process that creates the geometric effects Einstein described. Relativity describes what happens; substrate mechanics explains how it happens.

Quantum Mathematics and Substrate Waves

Quantum mechanics provides incredibly precise mathematical tools for predicting atomic and subatomic behavior. The substrate model suggests why these mathematical tools work: they're describing wave mechanics in the substrate medium.

The Schrödinger equation would remain the best tool for calculating quantum phenomena. But we'd understand that it's describing pressure wave propagation in a physical medium rather than mysterious "probability waves" in abstract mathematical space.

Particle Physics and Substrate Extremes The Standard Model successfully catalogues and predicts the behavior of fundamental particles. The substrate model suggests why particles have the properties they do: they're stable wave patterns in the substrate medium, with different particles representing different types of substrate displacement and circulation.

Particle accelerator calculations would remain accurate. But we'd understand that we're studying extreme substrate displacement effects rather than truly fundamental building blocks of matter.

The Hierarchy of Understanding

The substrate model suggests that physics operates at multiple levels of understanding, each valid and useful in its own context:

Level 1: Phenomenological (What We Observe) This is the level of direct observation - objects fall, magnets attract, atoms emit light. This level is completely reliable and forms the foundation of all science.

Level 2: Mathematical (How We Calculate)
This is the level of current physics - differential equations, field theories, quantum mechanics. These mathematical frameworks predict observations with extraordinary precision. They tell us how to calculate what will happen in any given situation.

Level 3: Mechanical (Why It Actually Works) This is the level the substrate model potentially addresses - the actual physical mechanisms that cause the mathematical relationships to hold true. It's the deepest level of understanding, explaining why mathematics works so well.

Each level is valuable and necessary. You don't need to understand substrate mechanics to calculate orbital trajectories or design quantum computers, just as you don't need to understand combustion chemistry to drive a car. But the deeper understanding enriches and illuminates the practical applications.

Cosmic Implications

If the substrate model is correct, it transforms our understanding of the universe on the largest possible scales.

The Dark Matter Mystery Current cosmology requires vast amounts of invisible "dark matter" to explain galactic rotation curves and cosmic structure formation. But if gravity operates through substrate pressure gradients rather than direct mass attraction, then these observations might be explained through substrate mechanics rather than exotic matter.

Galaxies might be organized by substrate wave interference patterns created by supermassive black holes. The complex rotation curves we observe might reflect substrate flow patterns rather than the gravitational effects of invisible matter. Dark matter might not exist at all - it might be a misinterpretation of substrate dynamics.

The Dark Energy Puzzle Similarly, the accelerating expansion of the universe currently requires "dark energy" - a mysterious force that pushes galaxies apart. But substrate wave mechanics might provide an alternative explanation.

If the universe is filled with substrate waves propagating at tremendous speeds, these waves might create large-scale pressure effects that influence cosmic expansion. The apparent acceleration might reflect substrate wave interactions on cosmic scales rather than a mysterious repulsive energy.

The Big Bang and Substrate Creation The Big Bang itself might be understood as the ultimate substrate creation event - a moment when vast amounts of substrate medium were created and began expanding to fill space. Instead of matter and energy expanding into empty space, we might have had substrate medium expanding and creating the foundation for matter and energy to exist.

This could explain why the universe appears fine-tuned for physics - it's not that the constants of nature happen to have the right values, but that the substrate medium has specific mechanical properties that determine what we observe as natural constants.

The Profound Implications

Understanding the universe as substrate mechanics rather than mysterious field interactions would have profound philosophical and practical implications:

The Return of Intuitive Physics For the first time in over a century, physics could become intuitively comprehensible again. Instead of accepting mathematical abstractions, we could visualize and understand the actual mechanical processes that create all natural phenomena.

The Technological Revolution If we can learn to detect and manipulate substrate effects, the technological possibilities become almost limitless. Gravity control, instantaneous communication, clean nuclear energy, programmable matter - technologies that seem impossible today might become routine engineering challenges.

The Spiritual Dimension Perhaps most profoundly, the substrate model suggests that we're not isolated beings in empty space, but integral parts of a universe filled with a dynamic, responsive medium. Every movement, every thought, every heartbeat creates patterns in the substrate that connect us to the cosmos.

We're not separate from the universe - we're expressions of it, temporary patterns in the same medium that creates stars, galaxies, and the forces that bind atoms together.

What We Still Don't Know

Despite the elegance and explanatory power of the substrate model, enormous questions remain:

Substrate Properties We still know almost nothing about the substrate's detailed properties. What is its exact density? How does its elasticity vary with temperature or pressure? How does it interact with different types of matter? These basic questions would require decades of careful experimental work to answer.

Detection Methods We need to develop reliable methods for detecting substrate effects directly rather than inferring them from their consequences. This might require entirely new types of instruments and measurement techniques.

Substrate Origins Where does the substrate come from? How did it come to exist? Is it created and destroyed, or is it eternal? These questions touch on the deepest mysteries of existence itself.

Biological Interactions How do living systems interact with the substrate? Does consciousness involve substrate patterns? Could substrate manipulation affect biological processes? These questions bridge physics and biology in ways we're only beginning to explore.

Technological Limits Even if substrate manipulation is possible, what are its practical limitations? How much energy would be required? What are the safety considerations? How long would it take to develop practical applications?

The Experimental Challenge

The ultimate test of the substrate model lies in careful experimentation. Theoretical elegance and explanatory power are important, but physics is ultimately about understanding the real world through observation and measurement.

The substrate model makes specific, testable predictions:

Characteristic oscillation signatures in high-energy particle collisions.

Substrate wave propagation at speeds far exceeding light.

Gravitational effects that deviate slightly from general relativity predictions.

Electromagnetic phenomena that show substrate circulation signatures.

These predictions can be tested with existing or near-future technology. The results will determine whether the substrate model represents a genuine advance in understanding or an elegant misconception.

A Humble Revolution

It's crucial to emphasize that the substrate model remains highly speculative. It might be wrong in its details, wrong in its approach, or wrong entirely. Science has a long history of promising theories that ultimately failed experimental tests.

But even if the substrate model proves incorrect, the questions it raises are important. Why do the mathematical frameworks of modern physics work so well? What is the actual mechanical reality underlying quantum mechanics, relativity, and particle physics? How can we develop a more intuitive understanding of natural phenomena?

These questions are worth pursuing regardless of whether the substrate model provides the correct answers.

The Wonder Continues

Perhaps the most beautiful aspect of the substrate model is how it transforms mystery into wonder without diminishing either. Understanding lightning as an electrical discharge doesn't make storms less awe-inspiring. Understanding how birds fly doesn't make flight less miraculous. Understanding how the substrate creates forces doesn't make the universe less magnificent.

If anything, the substrate model makes the universe more wonderful by revealing the elegant simplicity underlying apparent complexity. Instead of four separate, mysterious forces, we might have one comprehensible mechanism operating at different scales. Instead of incomprehensible quantum weirdness, we might have beautiful wave mechanics in the most subtle medium ever discovered.

The universe would still be vast, mysterious, and awe-inspiring. But it might also be comprehensible, mechanical, and inviting to human understanding in ways that current physics cannot provide.

The Journey Forward

Whether or not the substrate model proves correct, it represents something important: the continuing human quest to understand our place in the cosmos through direct, mechanical comprehension rather than abstract mathematical description.

This quest has driven every great advance in human knowledge. From ancient astronomers tracking planetary motions to modern physicists probing the structure of atoms, we've always sought to understand not just what happens, but how and why it happens.

The substrate model continues this tradition by asking: What if the universe operates through mechanisms as comprehensible as any human machine? What if physics describes an elegant engineering project operating on the cosmic scale? What if we're not isolated observers of mysterious phenomena, but participants in the most beautiful mechanical system ever created?

These questions inspire wonder, drive investigation, and open possibilities that seemed closed forever. They remind us that science is not about accepting limitations on human understanding, but about continuously expanding the boundaries of what we can comprehend and potentially control.

The substrate universe awaits our investigation. Whether it exists as described, exists in some modified form, or doesn't exist at all, the search itself advances human knowledge and capabilities. And in that search, we discover not just the nature of the universe, but the nature of ourselves as beings capable of understanding the cosmos we inhabit. The greatest adventure in human history - understanding reality itself - continues.

"The most incomprehensible thing about the universe is that it is comprehensible."

Einstein

EPILOGUE

THE READER'S JOURNEY

You've now taken a complete journey through the substrate universe - from the fundamental contradiction of attributing properties to nothingness, through the mechanical explanations of all known forces, to the technological and philosophical implications of substrate physics.

Whether you're convinced by these ideas, skeptical of them, or somewhere in between, you've engaged with some of the deepest questions humans can ask: What is the universe made of? How does it work? What is our place within it?

The substrate model offers one possible set of answers - elegant, mechanical, and potentially revolutionary. But the real value lies not in accepting these answers, but in continuing to ask the questions. Science advances through bold hypotheses tested against careful observation. The substrate model represents one such hypothesis.

The universe is still out there, waiting to reveal its secrets to anyone curious enough to look, question, and wonder. The journey of discovery never ends - it only gets more interesting.

The Nothing Problem - Comprehensive Glossary

Core Concepts

Aether - Historical term for a proposed medium filling space to carry electromagnetic waves. Modern substrate theory differs from classical aether models by proposing a completely dynamic, co-moving medium rather than a static or partially dragged one.

Cavitation - The formation of voids or low-pressure regions in a medium. In substrate mechanics, extreme matter compression creates cavitation zones where substrate density approaches zero, generating nuclear-scale forces.

Circulation Patterns - Rotational flow patterns in the substrate medium created by moving electric charges. These whirlpool-like disturbances in the substrate manifest as magnetic fields and electromagnetic interactions.

Nothing Problem - The fundamental logical contradiction in modern physics: space is described as "nothing" (empty vacuum) yet possesses measurable properties like wave propagation, energy storage, and field support capabilities.

Pressure Gradients - Variations in substrate pressure from point to point in space. These gradients drive acceleration according to the fundamental UDT equation and create what we experience as gravitational, electromagnetic, and nuclear forces.

Quantum Substrate (QS) - The hypothetical medium filling all space with measurable density ($\sim 1.0 \times 10^{-25}$ kg/m^3) and mechanical properties. This medium's displacement and flow patterns create all fundamental forces through classical fluid dynamics.

Substrate Displacement - The process by which matter pushes aside or displaces the quantum substrate, creating regions of altered substrate density and associated pressure gradients.

Substrate Medium - Alternative term for quantum substrate, emphasizing its role as the physical medium through which all forces propagate via mechanical interactions.

Unified Displacement Theory (UDT) - The theoretical framework proposing that all fundamental forces emerge from mechanical processes (displacement, pressure gradients, circulation) in the quantum substrate medium.

Technical Parameters

β (Beta) - Substrate displacement coefficient ($\beta = 3.333 \times 10^{-6}\ m^3/kg$) that quantifies how much substrate volume is displaced per unit mass of matter.

φ (Phi) - Substrate displacement parameter indicating the degree of substrate depletion in a region. Values range from $\varphi \approx 1$ (minimal displacement for gravity) to $\varphi \approx 10^{-30}$ (extreme cavitation for nuclear forces).

ρ_QS - Quantum substrate density with background value $\rho\_QS = 1.0 \times 10^{-25}$ kg/m^3. This parameter appears in the denominator of the fundamental acceleration equation.

∇P_QS - Substrate pressure gradient (mathematical notation for spatial variation in substrate pressure). The driving force behind all accelerations in UDT.

A - Universal substrate constant ($A = 6.67 \times 10^{-36}\ s^{-2}$) relating particle density to substrate displacement effects.

α (Alpha) - Fine structure constant ($\alpha = 1/137.036$) reinterpreted as substrate particle orientation efficiency ratio rather than arbitrary fundamental constant.

Physics Regimes

Cavitation Regime - High-density regions (nuclear forces) where matter compression creates actual voids or near-voids in the substrate, leading to extreme pressure gradients and strong forces.

Circulation Regime - Electromagnetic domain where moving charges create organized rotation patterns in the substrate medium, generating magnetic fields and electric forces.

Enhancement Function (η) - Mathematical relationship $\eta = 1/\varphi$ showing how substrate depletion amplifies forces. Higher depletion (lower φ) creates stronger force enhancement.

Pressure Gradient Regime - Gravitational domain with gentle, long-range pressure variations in the substrate causing relatively weak but universal attractive effects.

Relativistic Regime - High-velocity conditions where substrate rearrangement becomes increasingly difficult, potentially explaining why massive objects approach but cannot exceed light speed.

Force Mechanisms

Boundary Pressure - Higher substrate pressure at the edges of displacement zones that squeezes objects toward regions of substrate cavitation. The active driving mechanism in gravitational attraction.

Electromagnetic Coupling - Interaction between charged particles mediated by substrate circulation patterns, following electromagnetic field equations as the mathematical limit of substrate fluid mechanics.

Gravitational Coupling - Interaction between masses mediated by substrate pressure gradients, with coupling strength varying by material composition ($A_{rocky} \neq A_{gas} \neq A_{ice}$).

Nuclear Binding - Extremely strong forces generated when multiple particles create overlapping cavitation zones, with substrate pressure gradients at void boundaries providing the binding mechanism.

Substrate Viscosity - Resistance of the quantum substrate to rapid flow changes, explaining inertial effects and providing mechanical foundation for Newton's laws of motion.

Mathematical Framework

Acceleration Equation - The fundamental UDT relationship:

$a = -(1/\rho_{QS}) \times \nabla P_{QS}$, stating that acceleration results from substrate pressure gradients divided by substrate density.

Displacement Field (φ) - Scalar field describing local substrate depletion levels throughout space, with $\varphi = 1$ representing undisturbed substrate and $\varphi \to 0$ representing complete cavitation.

Enhancement Amplification - The mathematical principle that force strength scales as $\eta = 1/\varphi$, explaining why nuclear forces are ~10^{35} times stronger than gravitational forces.

Lagrangian Formulation - Mathematical framework expressing UDT using energy methods: $L = ½\rho\_QS(\partial\varphi/\partial t)^2 - ½k(\nabla\varphi)^2$ + interaction terms.

Substrate Field Equations - Wave equations governing substrate dynamics: $\Box\varphi + m^2\varphi = J(x,t)$, where J represents particle-induced sources.

Advanced Concepts

Black Hole Singularity - Region where substrate displacement reaches $\varphi \to 0$, creating complete voids where substrate-mediated forces cannot operate, representing breakdown of the medium rather than infinite curvature.

Hawking Radiation - Particle emission from black holes explained through substrate reconnection processes, where the medium attempts to repair discontinuities at the event horizon.

Magic Numbers - Empirically observed nuclear stability patterns reinterpreted as substrate flow diagnostics rather than universal constants, indicating local medium conditions.

Substrate Reconnection - Process analogous to magnetic reconnection where torn substrate patterns repair themselves, potentially releasing energy as particle-antiparticle pairs.

Substrate Waves - Mechanical waves propagating through the quantum substrate medium itself at estimated speeds of ~3.6×10^{34} m/s, potentially enabling faster-than-light information transfer.

Substrate Cymatics - Vibrational patterns in the substrate medium that may maintain atomic and molecular structures through resonance effects, similar to sand patterns on vibrating plates.

Technological Applications

Gravity Control - Hypothetical technology manipulating substrate pressure gradients to create artificial gravitational effects by generating controlled displacement zones.

Inertial Dampening - Proposed method of reducing inertial forces by manipulating local substrate density around objects undergoing acceleration.

Substrate Communication - Theoretical faster-than-light communication using substrate wave patterns rather than electromagnetic signals.

Substrate Engineering - Speculative field involving direct manipulation of substrate properties to control fundamental forces and create new technological capabilities.

Historical Context

Einstein's Speed Limit - The light-speed constraint that applies to matter and energy but may not restrict substrate wave propagation, allowing faster-than-light information transfer without violating relativity.

Green Paint Problem - Analogy illustrating how mathematical success doesn't negate underlying mechanisms, comparing substrate mechanics to Einstein's equations as green paint to yellow-blue mixture.

Michelson-Morley Experiment - Historical experiment that failed to detect aether wind, reinterpreted as evidence for co-moving substrate rather than proof against any medium.

Tesla Connection - Nikola Tesla's experiments with wireless power transmission may have been early observations of substrate wave effects, predating theoretical understanding by over a century.

Measurement Units

GeV (Giga-electron Volts) - Energy unit used in particle physics, particularly relevant to the predicted 2.4 GeV oscillation signature that could validate UDT.

kg/m³ - Standard density units for substrate medium ($\sim 10^{-25}$ kg/m³) and various matter configurations across different physics regimes.

Pa^{-1} (Inverse Pascals) - Compressibility units for substrate elastic properties ($\beta = 1.11 \times 10^{8}$ Pa^{-1}).

s^{-2} (Inverse Seconds Squared) - Units for the universal substrate constant $A = 6.67 \times 10^{-36}$ s^{-2}, relating to substrate response characteristics.

Experimental Validation

2.4 GeV Oscillation - Predicted energy signature in particle collision data that could provide experimental evidence for substrate mechanics, based on theoretical substrate resonance frequencies.

ATLAS Data - Particle accelerator experimental results analyzed for substrate-predicted oscillation patterns, showing preliminary correlation with UDT theoretical predictions.

Cross-Scale Validation - Testing substrate mechanics across multiple scales from planetary motion (gravitational regime) to particle collisions (nuclear regime).

Statistical Significance - Measure of experimental validation strength, with $\geq 5\sigma$ discovery-level confirmation needed for scientific acceptance of substrate mechanics.

Note: All entries represent theoretical concepts within the UDT framework. The speculative nature of these ideas requires extensive experimental validation before scientific acceptance.

The Nothing Problem - Conept Guide

Aether — Historical idea of a medium filling space; revisited through substrate mechanics in modern form (Ch. 2–3)

Acceleration — Explained as response to pressure differentials within the substrate (Ch. 4)

Black Holes — Described as regions of extreme substrate displacement and pressure inversion (Ch. 13)

Consciousness — Suggested to interact with substrate organization and reality formation (Ch. 15–18)

Displacement — The foundational act of matter interacting with the substrate, source of all forces (Ch. 1, 4)

Electromagnetism — Interpreted as circulation and motion of pressure waves within the substrate (Ch. 6)

Energy — Defined as organized substrate motion or stored pressure potential (Ch. 4–9)

Einstein — Referenced for relativity's link between spacetime curvature and substrate pressure (Ch. 1, 4, 5)

Field — Reinterpreted as pressure and flow patterns in the substrate medium (Ch. 3–4)

Force — Unified view: all forces arise from substrate pressure gradients (Ch. 4)

Gravity — Modeled as gentle substrate pressure gradients around displaced mass (Ch. 5)

Inertia — Understood as resistance to substrate acceleration; result of momentum exchange (Ch. 9)

Light — Wave motion through substrate, not through "nothingness"; defines speed and medium stiffness (Ch. 1, 6)

Mass — Expression of localized substrate density; interacts through displacement (Ch. 4–5)

Mechanics — Restored to physics through tangible substrate processes (Throughout)

Michelson-Morley Experiment — Historical test that disproved static aether, not dynamic substrate (Ch. 2)

Nuclear Forces — Extreme compression and pressure confinement within substrate at atomic scales (Ch. 7)

Pressure — Central mechanism; gradients produce gravity, electromagnetism, and nuclear cohesion (Ch. 4–7)

Quantum Connection — Substrate fluctuations underlying quantum behavior (Ch. 12)

Relativity — Described as emergent from substrate density variations and motion (Ch. 4–5, 11)

Space — Redefined as filled with a dynamic substrate rather than void (Ch. 1, 3)

Spacetime — Result of large-scale substrate flow and deformation (Ch. 4, 5, 13)

Substrate — The universal medium generating all physical effects; foundation of Unified Displacement Theory (Throughout)

Technology — Discussed as future engineering of substrate forces (Ch. 16)

Unified Displacement Theory (UDT) — Framework unifying all forces via substrate pressure gradients (Ch. 4, 18)

Universe — Interpreted as continuous dynamic substrate structure without emptiness (Ch. 15–18)

Vibration — Fundamental mode of energy and wave transmission in the substrate (Ch. 8)

Waveforms — Patterns of motion in the substrate forming light, energy, and matter (Ch. 6–8)

Author's Note: Why I Wrote This

As a teenager, I was captivated by stories of physicists searching for the "theory of everything" — a single framework that could explain all the fundamental forces. There was something almost magical about the idea that the universe might operate according to one elegant set of principles. But life took me in a different direction.

In 1999, I began a career in law enforcement, and for years, the dream of unified physics faded into the background. Then, in 2005, I became an accident reconstructionist, and physics suddenly became practical and immediate. I was using momentum calculations to understand collisions, analyzing trajectory paths to reconstruct events, and applying the very equations I'd once found mysterious to solve real-world problems.

But this wasn't the abstract, incomprehensible physics I'd encountered in textbooks. This was physics I could visualize, understand, and apply. It had to make mechanical sense because people's lives depended on getting it right.

My creative nature led me to explore these concepts through storytelling. I wrote and directed a short film called "Impulse" — a story about a man working backwards through the physics of his planned suicide, grappling with loss and the mechanics of tragedy. The film won awards, but more importantly, it showed me how physics and human experience could be woven together into something comprehensible and emotionally resonant.

The real breakthrough came during the OceanGate tragedy in 2023. I became obsessed with understanding the hydrostatic pressure that had caused the submersible's implosion. As I worked through the mathematics of pressure gradients surrounding the vessel, something clicked. The pressure wasn't just pushing inward — it was creating a dynamic flow pattern around the entire structure.

Suddenly, I saw the connection. What if gravity worked the same way? What if massive objects like planets created pressure gradients in some medium, and what we experience as gravitational attraction is actually pressure pushing us toward areas of lower density?

That moment changed everything. The pins started falling into place. If gravity was fluid dynamics, what about other forces? What about the medium itself? What if space wasn't empty but filled with something so subtle we'd never directly detected it?

The teenager who'd dreamed of a unified theory was still there, and he finally had a path forward.

When I began developing Unified Displacement Theory, it started as a mathematical exercise — could the fundamental forces be unified through a single substrate field equation? But as mathematics developed, something remarkable happened: the abstract formalism began revealing mechanical processes. The equations weren't just predicting observations; they were describing how things actually work.

For the first time in years, physics felt comprehensible again.

I found myself able to visualize gravitational effects as pressure gradients, electromagnetic forces as circulation patterns, and quantum mechanics as wave phenomena in a physical medium. Mathematics became more than calculation tools — they became windows into an elegant mechanical reality underlying all natural phenomena.

But here lay a dilemma: the technical papers, with their mathematical rigor and statistical analyses, could communicate these insights only to fellow researchers.

The profound shift in understanding — from mystery to mechanism, from abstraction to intuition — remained trapped within mathematical formalism.

I realized that the most important aspect of this work wasn't the equations themselves, but what they revealed about the nature of understanding. If physics could be made mechanically comprehensible without sacrificing accuracy, then we had an obligation to make it comprehensible. Not just for scientists and engineers, but for anyone curious about how reality actually works.

We live in a world where scientific literacy is crucial for informed citizenship, yet science itself has become increasingly incomprehensible to the citizens it serves. People are asked to trust scientific authority while being told that scientific truth is fundamentally beyond their comprehension. This creates a dangerous dynamic where science becomes dogma rather than understanding.

I wrote this book because I believe we can do better. I believe physics can be both mathematically rigorous and intuitively satisfying. I believe that anyone willing to think carefully about how things work deserves access to genuine understanding, not just simplified analogies or abstract formulas.

Most importantly, I believe that making physics comprehensible doesn't diminish its profundity — it reveals depths of elegance and beauty that mathematical abstraction alone can never fully capture. When you understand that you're swimming in an invisible ocean of substrate, that every force is a flow in this medium, that every molecule is held together by resonant oscillations — the universe doesn't become smaller or less mysterious. It becomes more magnificent because it becomes comprehensible.

This book represents my attempt to build a bridge from the technical mathematical framework to genuine human understanding. Not everyone needs to master differential equations to appreciate the elegance of nature's engineering. Not everyone needs to become a physicist to contribute to our understanding of reality. But everyone deserves the opportunity to grasp how the universe actually works, if they're curious enough to look.

The questions explored in these pages — What is space? How do forces work? Why does matter exist? — belong to all of us. They're human questions, not just scientific ones. And if there are mechanical answers to these questions, then those answers should be accessible to anyone willing to think clearly about them.

Whether the substrate model proves correct or not, the exercise of seeking mechanical understanding has value in itself. It trains us to ask deeper questions, to demand coherent explanations, and to trust our intuitive sense that the universe should make sense. These are habits of mind we need more of, in science and in life.

My hope is that this book inspires some readers to explore the technical aspects of substrate physics further. But my deeper hope is that it inspires all readers to reject the false choice between mathematical sophistication and genuine understanding. The universe is both mathematically elegant and mechanically comprehensible. We can have — and deserve — both.

My background in law enforcement has taught me that theories are only as good as the evidence behind them. I'm not claiming to have solved the mystery of the universe — but I am saying this:

"Probable Cause exists in a substrate medium theory — let's finish the investigation."

Kris Michael McKenna
July 2025

Made in the USA
Monee, IL
04 November 2026

67387617R00115